Patrick Moore's Practical Astronomy Series

The Complete CD Guide to the Universe

Practical Astronomy

Richard Harshaw

With 31 Figures

 Springer

Richard Harshaw
Stardeck Observatory, Kansas City, MO, USA
1817 NE 83rd Street
Kansas City 64118
Email, personal: dwharshaw@kc.rr.com

British Library Cataloguing in Publication Data
A catalogue record for this book is available from the British Library

Patrick Moore's Practical Astronomy Series

Additional material to this book can be downloaded from http://extras.springer.com

ISBN-10: 1-4899-9738-5   ISBN-10: 0-387-46895-1 (eBook)
ISBN-13: 978-1-4899-9738-8   ISBN-13: 978-0-387-46895-2 (eBook)

Printed on acid-free paper

Printed in the United States of America (TB/EB)

9 8 7 6 5 4 3 2 1

Springer Science+Business Media
springer.com

To my best friend and constant companion, my wife Loretta, who indulges my crazy pursuit of one of the greatest hobbies in the world. She puts up with odd hours, strange travel plans, an unfathomable pursuit of gadgets and gizmos, and an incomprehensible vocabulary of stellar and galactic terminology. Yet she always has a gentle way of bringing me back to earth and taking the time to enjoy the simple daily pleasures of life too.

To my granddaughter, Alexis. Her five-year-old curiosity and endless parade of questions sometimes wears Grandpa down, but never do I tire of her desire to know. I hope the day comes when she can take up the observations of the skies over her head on her own. Until then, I am here to guide you, sweetheart. I love you more than two hundred billion galaxies!

Contents

Preface

If you are looking for a book full of "eye candy"—stunning celestial images that can inspire your imagination and induce a sense of awe at the grand scale of our marvelous universe—this book is not for you.

If you are looking for a good "coffee table" book to lay out on your living room furniture so guests can thumb through the beautiful pages and perhaps ask you questions about what the book contains—this book is not for you either.

But if you are looking for a book that can help you observe more of the heavens than you ever thought possible, then this book might be what your are looking for.

I wrote this book after I upgraded my telescope and found myself with a much larger universe to observe than I had before. How do I plan an observing program that can help me make the most of my limited time and squeeze the most out of my new piece of finely figured glass?

It soon dawned on me that observing the sky by constellation—which is a very popular option—was not, in the Spockian sense, "logical" to me as many rich parts of the sky spill over the artificial constellation boundaries people have assigned to the sky. (For example, the Winter Milky Way sprawls across several constellations and if you are pursuing, let us say, open clusters along the Milky Way's mid-plane, you may leave out several excellent views if you limit yourself to Canis Major, or Orion.) I decided to divide the sky into "zones" and study it by zone as the combination of seasons, weather, and lunar cycles permitted. The result is the format you will use in this book. (The astute reader will accuse me of substituting one set of man-made boundary lines for another, to which I reply.

True! You got me! But the zones I use are much smaller than the constellations so you can spend an entire evening in a small part of the sky and really enjoy all it has to offer at whatever pace you wish to use.

Indeed, some zones—which measure one hour of right ascension in width and 10 degrees of declination in height—may take you four or five nights, or more, to fully explore!)

All of the observations reported in this observing guide were made from suburban locations. Only a small number (less than 2%) were made from truly "dark sky sites," and those records will be noted when you encounter them.

From 1987 to 1990, I made my observations in Columbia, Missouri using a Celestron C-8. Columbia at that time was a city of about 100,000 people. My residence was located on the southwest edge of town which put most of the light dome from the city to my east and north. This was fortunate because a long, low ridge lay between my home and the city. This ridge was high enough to block out the majority of the truly offensive night glow, with only a small fringe of faint sky glow extending up to about 20 degrees above the horizon. But since I did not have a clear view of the eastern sky from my residence, I did not need to observe in that direction anyway. My views overhead, to the North, South, and West were clear with skies often having a naked eye limiting magnitude of 5.8 to 6.0.

In 1990, I moved to Kansas City, MO and built a home on the north side of the city. I was located about 12 miles due north of the center of the city, with a decent horizon all around me. But the awful sky glow from the millions of watts of high-pressure sodium street lighting was much more offensive than the modest glow of Columbia. So my southern sky below about −20° declination is almost always awash in a hopelessly bright glow. Skies to the East and West are better, the zenith good, and the North even better. On a good night from my location, the limited naked eye magnitude can reach as low as 5.5, but usually 4.0 or so on a typical night (at the zenith).

In 2000, I purchased a Celestron C-11 and have been observing with it ever since.

My main point to all of this is that *everything described in this book can be observed from suburban sites with instruments of moderate aperture.* Huge "light buckets" or wonderfully dark skies are not a requirement to detect the wonderful treasures described in this book. They help, of course, but the fact is that you can observe a lot of things in the sky from even brightly lit suburban sites. See the discussion on observing galaxies in Chapter 3.)

In this guide, you will have descriptions of 13,238 *objects* viewed from the sites I previously described. The majority of them—10,738—are double or multiple stars. The balance—2,500—are "deep sky" objects (as if double stars were not in the "deep sky"!).

Double stars dominate this work for several reasons. *First,* for the modest aperture telescope, there are far more of them than anything else in the sky. The Washington Double Star Catalog (or WDS as amateurs often call it) is considered to be the standard double star reference in the business today. It lists well over 100,000 pairs. If we filter the WDS and remove from it those pairs that are (a) too faint to see in scopes of 11-inches or less aperture, (b) too close to separate in such instruments, and (c) too far south to be seen from the north 40th parallel, we end up with about 20,000 pairs. I have chosen 10,596 of the best and have not included another 8000 or so pairs I have observed that are just too faint or difficult to be of much interest to a general observer. As much as possible, I have tried to include only true binary stars, not just chance alignments of two stars that happen to "look" close together. (For that reason, many of the popular "double" stars you may see on some lists are not included in this book.) Where a pair is in doubt, I will make remarks to that effect.

Second, double stars are usually bright enough to be easily observed from even urban sites. Double stars can often be seen in hazy, murky, moonlit skies, and in types of weather that render galaxies, planetary nebulae, and other faint and extended objects simply invisible.

The deep sky objects include many galaxies (1573 of them), and the point to be made is that although dark sky sites help in tracking down and bagging these elusively faint blotches of light, many of them can still be observed in what most amateurs would write off as hopeless skies for the task. True, dark skies reveal more detail in galaxies than suburban skies, but do not think that suburban skies mean your galaxy-hunting efforts will be limited to a handful of bright Messier objects!

You will also find 580 open clusters at your disposal, 109 globular clusters, 148 planetary nebulae, and a handful of other interstellar and intergalactic stuff.

If you are an urban or suburban amateur with limited observing time and want to get the most out of your telescope, let this book guide you to all the treasures that are within your grasp. I promise you a rich and rewarding journey and memories to last more than a lifetime!

Richard Harshaw
Kansas City, MO

Foreword

This is a book and CD-ROM for people with a telescope and who are interested in an organized study of the wondrous night sky, including its double stars, clusters, nebulae, and galaxies. While there are other books covering these topics in the amateur literature, this one is better suited to the amateur astronomer wanting to observe the sky with a systematic system and keep track of their observations.

This work starts out with the author recalling his early interest in astronomy (he has been watching the night sky for over 40 years) and works up to the topics that give a better understanding of the book and how to get more out of observing. There are some recommendations for supplemental reading if needed. Some important basics are covered to give the reader a foundation for the remainder of the book and CD-ROM.

A major theme of this book (and of the author's lifelong observing pursuits) is double stars. There is information on what measurements apply to a double star and how they are made, stellar color, and the dynamic nature of binary systems. Included is a helpful and detailed discussion on the naming and "coding" of double stars. There are biographical thumbnails of some of the more important double star observers—a very interesting historical footnote! Finally, there is a discussion about the use of scale models for doubles in a kinetic attempt to help the observer get a feel for what he or she is observing when viewing a binary star system.

But there are also a rich host of deep sky objects listed throughout the book. Here you will find an explanation of the nomenclature of galaxies, open clusters, globular clusters, and planetary nebulae.

After all of this introductory material, the real meat of the book begins—how to use the CD-ROM. There are, altogether, about 13,000 pages of material on the CD and a little time and care spent early on learning how to navigate it will save the reader a great deal of time later.

The heart of the system is in the finder charts and observing catalogs. There are four sections of maps and catalogs, one for each season. Each seasonal section is then divided into zones for each hour of *right ascension* and 10 degrees of *declination*, down to −40 degrees.

It is in the zones where you will find the most interesting data. Each zone, in Adobe Acrobat (PDF) format includes detailed maps covering its own area of sky. (For those who do not have Adobe Acrobat Reader installed on their computers, the CD has the Reader installation program on it.) A unique feature of Richard's work is that all maps (except the Zone Master Map) are given twice—one map being a "normal" image and the other a "mirror" image for use at the telescope. For those who have ever tried to mentally reverse or invert a field from a finder chart while at the telescope, the sheer ease and utility of the mirror image charts should bring a tear of joy to your eyes. (Just do not let that tear drop onto your eyepiece . . . .)

Each zone has a detailed index. The index lists all the objects that are listed for that zone in alphabetical order, with references to which detailed finder chart in which the observer can find that object.

Last are the objects themselves. The double stars are listed first, followed by deep sky entries. In addition to name and location, other details about the object are given along with Richard's own observation notes for each object. (Yes, this is one observing guide where the writer has actually observed every object in the guide!)

Objects are presented in a unique way to facilitate observing and develop your observing skills. Easy objects (both double stars and deep sky) are listed first, followed by objects of moderate difficulty and culminating with difficult objects. Within each difficulty class, Richard has also ranked the objects in order of the eyepiece impact, with stunning objects being listed ahead of moderate ones, which in turn are listed ahead of those all-too-unimpressive views we often see as observers. There is great utility in this—one can decide on a night of moderate to poor seeing to focus on the easy objects and save the difficult ones for a night of better seeing. (This will especially be true of close and difficult double stars.)

Richard has also prepared a set of PDF files of objects by constellation for those who prefer to hunt the sky by constellation. As in the Zone format, each starts with easy double stars, works toward the more difficult, and then covers deep sky objects. The documents can be printed and used to keep track of observations.

One of the more practical things about this book is that every object described in it has been observed from suburban skies, with only a few exceptions (and these are noted in the text when they occur). Too often in the past, I have seen observing guides that require skies that frankly are just not available to the majority of amateur astronomers in the world, and Richard's work says to all who will try, "these things can be seen with a modest scope (8 inches or large) from even well-lit suburbs!"

I have observed with Richard on numerous occasions when his travels bring him to Phoenix, AZ, and can attest to his experience. But what impresses me the most about Richard is his enthusiasm for the night sky and his willingness to share it with any who will listen. One night at our annual Sentinel Star Gaze, for instance, I was trying to locate one of the globular clusters in M31 with my 14.5-inch Dobsonian. Richard had observed it only a couple of months earlier in his C-11 from skies that were much more difficult to look through than our wonderful desert skies, and he helped me hunt down the illusive little beast. When I got to confirm my first view of a non-Milky Way globular, Richard was just as excited about my conquest as I was! I am sure that you will find that same spirit of excitement and adventure in the book you are about to take home.

As Richard would say, "Clear skies, and fully dilated pupils to you!"

A. J. Crayon
Phoenix, AZ.

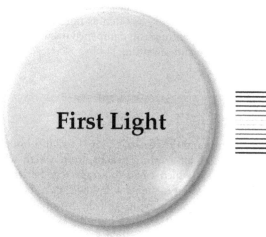

First Light

It was his first look ever through a telescope, and he could hardly contain his excitement as he swung the tube on its alt-azimuth mount and pointed it in the direction of the first quarter moon. He locked the axis screws and peered through the finder scope. (He had spent part of the afternoon aligning it on a church steeple a mile away.) There, just a little below and to the left of the cross hairs, was the quarter moon! He unlocked the set screws and clumsily centered the cross hairs on the target and then—with the expectancy of a little child at Christmas—he looked into the eyepiece.

Words could not describe how that 14-year-old boy felt! There were craters, just like his astronomy book showed! And mountains, and maria! The jagged rip of the terminator was a source of incredible fascination to him.

After several minutes of absorbing his first telescopic views of the moon, he decided to try to conquer more ambitious targets. Being the late spring of the year, the youngster loosened the set screws and slewed the tube eastward to the vicinity of the Keystone in Hercules, low in the eastern sky. He wanted to look at the great globular star cluster, M13. (He was not sure what the M stood for, but the pictures of M13 in his astronomy book were breathtaking.) He pointed the tube at the spot on the western side of the Keystone where his little star chart said M13 was, and used his finder scope to fine tune his aim. When he thought he was in the right area, he went to the eyepiece and—no M13! Patiently, he worked his way back and forth in the field until suddenly a fuzzy ball of light emerged into view like a ghost on a foggy moor. "Is *that* it?" he mused to himself. "That is not how it looks in my book," he thought. Where were the myriad dots of light his picture showed? This was not much at all—just a big, round patch of ill-defined fog!

He then spent 20 minutes in a fruitless effort to find M87 in Virgo. He then took several minutes to get Alcor and Mizar in the Big Dipper's handle in the eyepiece before he could enjoy the first of over 21,000 "double stars."

And so it went, the rest of the evening (although not very late—the next day was a school day), trying to locate things he had seen in his astronomy book and looking where the star charts said to look, as often as not failing to locate the object, but nonetheless not finding his enthusiasm dampened one bit by his failures.

It was the start of a rich and rewarding hobby.

Aperture Fever

That youngster was me, and the year was 1965. I had just bought my first telescope, a 60-mm refractor on an alt-azimuth mount, from a high-school senior who sold it to me for a hefty $15 (which I had to scrape together from several months of allowances and chores), and it included a Barlow lens, a solar projection screen, and two 0.925-inch eyepieces, giving me magnifications of 30 × and 100 × (60 × and 200 × with the Barlow). The alt-azimuth mount was smooth, even if it could not be used to locate objects by their sky coordinates, and the tripod was a fairly sturdy one. Such a telescope today would probably cost around $300 new.

For the next 4 years, I used my little f-15 refractor to explore first the easy things (the moon, the planets, bright galaxies, star clusters, and easy double stars). Then, as my skills improved, I went for more and more difficult targets, eventually pushing the modest optics of my scope to their limits. I saw the polar caps of Mars, the Red Spot on mighty Jupiter (at that time, a light brick red, not the salmon pink of today), the Cassini Division in the rings of Saturn, the moon's Straight Wall and Alpine Valley. I spent many nights in awe gazing at the Pleiades or M11. I strained to make out the Crab Nebula and several faint galaxies. I even spent one bitterly cold winter night watching the moons of Jupiter go through their majestic, formal dance.

In my junior year at high school, I gave the refractor to a younger neighbor and bought my first "serious" scope—a 4.5-inch reflector with an equatorial mount. And thus began a new era of beginner's frustration!

I had never used an equatorial mount before but knew from my readings that if I ever hoped to easily locate the sky's more difficult (a euphemism for "faint") objects, I would find one helpful. The frustration came in learning what the sky's coordinates were all about, and then learning how to accurately "polar align" the thing so I could find those faint fuzzies the books all talked about. I will cover in later pages how I now accurately polar align, but suffice it to say for now that my last years of high school were times when I had a love–hate relationship with my reflector!

What I Assume You Know

At this point, I assume you already have a fairly good *basic* understanding of the elements of visual astronomy. I assume you

- know the difference between a refracting telescope and a reflecting telescope,
- know what is meant (roughly) by "polar aligning,"
- know the difference between an alt-azimuth and an equatorial mount,

- understand how to use an equatorial mount and that you are conversant with astronomy's system of longitude (right ascension) and latitude (declination),
- have a fundamental grasp of the magnitude scale,
- know the basic types of things that there are to see in the sky with a telescope,
- know what is meant by "aperture," and
- know how to compute the magnification of a telescope given the focal lengths of the objective and the oculars.

If these assumptions are beyond your present skill level, I suggest you first read any of the several excellent manuals on these topics for beginners, such as Phil Harrington's *Stare Ware*, David Eicher's *Beginner's Guide to Amateur Astronomy*, or Sam Brown's *All About Telescopes*. If you own a 20-cm (8-inch) Schmidt-Catadioptic telescope, as I do, you might also enjoy Peter Manly's *The 20-cm Schmidt-Cassegrain Telescope*.

Light Gathering Power

The 4.5-inch reflector offered me one great advantage over the 60-mm refractor—light gathering power, or in the vernacular of the amateurs, light grasp. I had learned that a manufacturer's claims about a telescope's magnifying power were secondary, and in most cases grossly exaggerated. The main thing an astronomer needs a telescope for is to gather photons, and for that the diameter of the *objective* is the governing factor. My 4.5-inch reflector had over 3.6 times the photon grasp of my refractor. Since each step of the Pogson or *magnitude* scale represents an increase of 2.54 times the light of the next level, this meant I could extend my reach a little over one full magnitude! Theoretically, my 4.5-inch reflector could reach magnitude 12.5 while the refractor could only reach down to 11.3. This meant, according to one of my astronomy books, an increase in the number of objects visible to me from about 870,000 with the refractor to 3,500,000 with the reflector.

The magnitude grasp of a telescope is given by the formula

$$M_F = 6.5 - 5(\log, \text{eye pupil diameter}) + 5(\log, \text{objective diameter})$$

where the eye pupil diameter and objective diameters are both measured in inches. If we use 0.20 inches for the typical dilated pupil (roughly 5 mm), we see that a 6-inch telescope, for example, should be able to reach all the way down to 13.9 magnitude.

Wow! In one fell swoop (at the cash register), I had almost quadrupled my visible universe!

I also knew that with increasing objective diameter came an increase in a telescope's ability to show detail—what astronomers call *resolving power*, or resolution. The simplified astronomy text I was using at the time gave a formula that I have since learned was a great simplification of the original *Rayleigh Formula*,[1] and it expressed a telescope's resolving power as its ability to discern details expressed in seconds of

---

[1] The Rayleigh limit formula for the resolving power of a telescope is given by $2.5 \times 10^5 (\lambda/d)$ where λ and d are expressed in meters. For optical values of λ, the formula approximates to $0.13/d$ arcseconds. The Dawes limit, determined by empirical measurements, yields $0.12/d$ arcseconds. Converting these to English units of measurement shows that where Dawes shows $4.5/d$, Rayleigh shows $4.87/d$.

Table 1.1. Couteau's Rules

r*	Effect
1.00	Stars separated
0.98	Stars touching tangentially
0.90	Stars form a figure "8"
0.85	*Flattened figure "8" (corresponds to the Dawes limit)*
0.80	Stars form a narrow rod
0.75	Stars form a rod
0.70	Stars form a rod
0.60	Stars form an egg or olive shape
0.50	Stars form a slightly oval image

*r, Rayleigh limit.

arc. The formula, in its simplest form, says to divide 4.5 by the telescope's objective diameter in inches. In the case of my old refractor, this would have meant 4.5/2.36 or 1.9 seconds of arc. In other words, I should have just been able to resolve (or, in the amateur's lingo, "split") a double star whose members were about 2 arcseconds apart or see details in planets and extended objects about 2 seconds in size. With my new reflector, the resolving power dropped to 1.1 seconds of arc, almost twice as good as the old refractor.

That is, until I learned what a reflector's secondary mirror and mounting system can do to the light path and resulting resolution. The secondary mount and mirror create ripples in the incoming light that opticians call interference. The net result is that the telescope cannot resolve quite as finely as the formula predicts. For me this meant that reflector had more light grasp than my old refractor, but not a whole lot better resolution. (With my 8-inch and 11-inch Schmidt Catadioptic Telescope (SCTs) the large central obstructions caused by the secondaries play havoc with stellar images under certain conditions.)

But all is not lost! The Rayleigh limit and Dawes's research show the *theoretical* boundaries for a telescope. In actual practice, most *good* telescopes can go surprisingly *beyond* the Rayleigh limit! My 8-inch SCT has a theoretical resolving limit of 0.57 seconds of arc, but after careful collimation (setting the optical path for perfect alignment), I have split stars that were closer than this. In fact, Paul Couteau, the prolific French double star observer, developed an empirical rule about how close pairs would look in a good telescope. Using "r" as the theoretical limit of resolution (the "Rayleigh limit"), Couteau's rules have been summarized in Table 1.1.

Recent experiments by a number of seasoned double star observers around the globe suggest that Couteau might be onto something with his "rules." I have a number of observations in my logs that suggest duplicity below the Dawes Limit!

At any rate, for the next 4 years, I explored the heavens with my reflector and honed my skills with the equatorial mount. This in turn allowed me to locate many objects that I had always wanted to see but never could locate. It was one of the most rewarding times in my life as an amateur astronomer.

During my freshman year at college, I decided to sell the reflector and start saving up for a really good telescope—something in the 10- to 12-inch range. But a series of unforeseen events forestalled the purchase of my dream scope until I was well out of college. And then I made a mistake—I bought another 60-mm refractor. At least, it had an equatorial mount.

I kept the second refractor for a couple of years and sold it too. This time I vowed to earnestly pursue my dream scope.

But it was to be 5 years before the opportunity to acquire it presented itself, and it involved an agreement I struck with my wife—we would split our income tax refund and each of us get something we really wanted. (She knew what that meant in my case!) I had picked out either an 8-inch SCT or a 10-inch Newtonian and was about to place my order when, through an unlikely chain of events that does not make a good story line, I obtained my 8-inch SCT for a fraction of the cost of a new one, buying it second-hand from a fellow only 40 miles from my home.

I have had that Celestron Classic C-8 for over 19 years now and have seen wonders that I thought I would never see. I also had to learn a whole new way to observe the heavens, since so many things were available to me, and so little time.

Then, in 2000, I added a Celestron C-11 to my arsenal, and, in 2001, housed it in an observatory in my back yard. The observatory is a roll-off roof type, and this set-up has allowed me to observe more in the last few years than all my observing time up to 2000!

CHAPTER TWO

Astronomical Mechanics

Modified Star Hopping

As you probably already know, the equatorial mount consists of two axes set at right angles to each other. One axis—the *polar* axis—points to the *north celestial pole* or that point in the sky where the earth's rotational axis points. The other axis—the *declination* axis—is allowed to rotate around the polar axis and permits pointing the telescope north and south. The beauty of this system is that once the polar axis is accurately aligned with the north celestial pole (that is, parallel to the earth's axis), the user only needs to rotate the telescope around the polar axis using a motorized drive system to compensate for earth's rotation and thus keep objects centered in the eyepiece for long periods of time.

Attached to each axis is a round graduated plate or collar known as a *setting circle*. Since there are two axes on an equatorial mount, there are two setting circles. Setting circles derive their names from the fact that because they are graduated with numbered scales, they can be used to "set" the telescope's aim by using the stellar coordinate system. Thus by turning the two axes until a certain right ascension (RA) and declination (dec) appear under the pointers on each axis scale, the telescope should be pointing at that coordinate's position in the sky. If that coordinate is for a galaxy or a double star, when the observer goes to the eyepiece, that object should be more or less centered and ready to observe.

Right ascension setting circles are movable. Depending on the manufacturer, they will be either snugly held in place by friction or locked in place by set screws. In either case, their movability is meant to allow the observer to (1) get a star of known coordinates in view in the eyepiece, and then (2) go to the setting circle and turn it until the pointers on the axes line up with the RA coordinates for the star. Then, (3) the telescope may be aimed at a point in the sky by unlocking the RA and dec axis brakes and turning the telescope until the new coordinates are under the pointers. The axis brakes can then be locked and—hopefully—the target will be in view when the user goes to the eyepiece.

Thus, using setting circles is a modified version of "star hopping." But it is far easier and more accurate than traditional star hopping. But beware of this: if you ever attend a star party, you will probably run into someone who sneers at you because you use setting circles instead of star hopping. "Why, you will *never* learn the sky that way," they will say. Each to his own. With my limited supply of observing time, I *do not* want to waste it hopping around in the sky in search of faint targets. I want to drive right up to the front door and get down to a visit! (And let us not open the can of worms of snide comments you may get if you attend a star party with one of the new GOTO telescopes . . . .)

Polar Alignment

The effective use of setting circles demands accurate alignment of the telescope's polar axis with the earth's axis. Polaris, the "Pole Star," does not sit exactly at the north celestial pole. It is about 40 minutes or so from the true pole, so if one were to align on Polaris, he or she would be off by almost a degree. And this error is more than the width of the field of view of most general purpose eyepieces.

I found a simple method of polar alignment that only takes a few minutes of time. I learned it from the technical experts at Orion Telescope Company of California.

To understand this method, look at a high quality star chart. You will notice that Polaris lies just east of the 2-hour meridian (2 hours, 45 minutes to be exact).[1]

This fact is the key to the Orion method. As it turns out, a good sampling of bright and easily accessible stars also lie on or near 0245, as well as 1445 (the same meridian extended through the north celestial pole and down the opposite side of the sky). We can use these stars and Polaris to walk a telescope mount's polar axis almost exactly to the north celestial pole.

Good target stars on the same side of the Pole as Polaris (i.e., on or near 0245) include ε Cassiopeia (at 0154+6346), γ Andromeda (0204+4222), and α Aries (0207+2331).

On the opposite side of the sky, you could use Arcturus (α Bootes, at 1416+4222), α Draco (1404+6420), or β Ursa Minor (1451+7413). As a rule, the farther from the north celestial pole, the better, though.

My personal favorites are Arcturus (for early spring/summer) and γ Andromeda (the rest of the year). Both stars are fairly distant from the north celestial pole and are also easy to identify in spotter scopes.

To align the polar axis on the north celestial pole, begin by setting up the telescope so that the tripod head is as level as possible. (I use a good quality bubble level for this.)

Next, if you have a clock drive, turn it on.

---

[1] For the remainder of this book, I will abbreviate coordinates like this: HHMM+DDMM or HHMM−DDMM, where the HH and MM are hours and minutes of RA and the DD and MM are degrees and minutes of Declination, + being north of the celestial equator, − being south. Thus Polaris's RA would be 0245. Its declination is +8918, so its full position using this shorthand would be 0245+8918.

Now, find the reference star you want to use and center it in your eyepiece. Set the telescope's setting circles for the coordinates of this reference star. If the declination is off, loosen the declination scale plate and rotate it to the star's declination and secure it.

Unlock the axis brakes and slew the telescope to Polaris's *coordinates* (not Polaris) and lock the brakes. If Polaris is not in the eyepiece, loosen the azimuth and altitude locks on the *mount* and rotate the mount east or west, north or south, or both, to center Polaris.

Next, go back to the reference star again and center it in the eyepiece and reset the setting circles to read the star's coordinates. Again, aim to Polaris's *coordinates* and see if it is centered in the eyepiece. If it is not, repeat the mount adjustment, then repeat the entire process until Polaris is centered and no more adjustments are necessary.[2]

Accurate Setting Circles

Many of the objects you will find in this book are incredibly faint or get lost in stunningly rich fields of the Milky Way. You will find your enjoyment of the sky greatly enhanced if you have accurate polar alignment and accurate setting circles.

To be as accurate as possible, the setting circles should be as large as possible (or, better yet, digital). For instance, my C-8 has a large diameter RA scale that is easy to aim to within 1 minute of arc. But the declination scales on the Celestron C-8 are woefully too small to be accurate. So I removed the declination circle from the right side of the fork and replaced it with an 8-inch diameter scale I obtained from the Oregon Rule Company in Portland, Oregon (web site is www.oregonruleco.com). Digital setting circles are even more accurate, and the ultimate would be the computer-assisted telescope! (My C-11 uses digital circles, and I can also connect it to my laptop computer and use it in conjunction with Software Bisque's powerful program, *TheSky*.)

If you go the high-tech route, be prepared for scoffers who will jab at you for "cheating" and not learning the sky. All I can say is that I own a telescope to see the sky, and I do not care how I get there as long as the wonders of the heavens can be found and I can have a nice visit once I arrive.

---

[2] This process is super for visual astronomy, but if you want to do astrophotography or want greater aiming accuracy, you will need to amore accurate alignment. For such accuracy, you will need an *illuminated reticule eyepiece*. Such an eyepiece has a graduated scale etched into it that can be illuminated softly by a red light-emitting diode (LED). Once the Orion method of alignment has been performed, insert the illuminated reticule into the eyepiece holder. If the eyepiece gives less than 150× of magnification, use a Barlow to get it above 150×. Find a star as close to the celestial equator as possible and as close to the meridian as you can (no more than 5° from the equator or the meridian). Center this star on the reticule's scale and watch the star for several minutes. If the image drifts *south*, the telescope points too far to the *east*. Loosen the mount's set screws and tap it lightly toward the west. Repeat the process until the drift stops. Conversely, if the star drifts *north*, the mount points too far to the west.

The final step in this extra-precise alignment method requires that you select a star within 5° of the celestial equator and as close to the eastern horizon as you can get and watch the drift again. This time, if the star drifts *south*, the polar axis points *below* the pole—raise it. Repeat the process until the drift stops. (Conversely, if the drift is *north*, you need to lower the polar axis.)

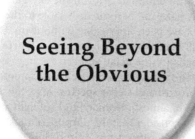

CHAPTER THREE

Seeing Beyond the Obvious

The Eyes Have It

The goal in astronomy is to gather photons. The aperture of the telescope is the chief factor in this process, but the eye of the observer is the other. A 400-inch telescope is useless if the observer is blind. I have known observers who could see more with a 4-inch telescope than those who owned 10-inch models. The secret is in how you use your eyes.

The human eye is not very efficient at gathering light. As a sensory receptor, it is designed to work well in daylight, not darkness. Also, the eye's sensitivity to light varies greatly from one person to the next.

If you have inherited poor night vision, there is not much you can do about it except compensate for it with aperture. (This is a costly cure!) But there are things you can do to help your night vision no matter how your genes are programmed.

For instance, it takes the average eye about 30 minutes to adapt to darkness, so begin your observing sessions by letting your eyes thoroughly "dark adapt."

This implies that the selection of observing sites is critical. A street light 50 yards away could make one as blind as the proverbial bat as far as seeing galaxies is concerned. Find as dark a site as possible. If you can afford it, build an observatory so you can totally block off stray ground-level light.

Another factor in site selection is the clarity of the sky. There is not much you can do about this if your area, like mine, is normally full of wind-blown dust and pollen, water vapor, and aerosols. Airborne contaminants reflect light from ground sources, giving the sky a ghostly glow. This glow very effectively blocks out the feeble light of the spiral arms of galaxies and extended nebulae.

Do Filters Help?

One partial remedy for sky glow is to use a light pollution-reducing filter. There are many on the market, and they go by a confusing array of names. Some are better than others. Good filters can eliminate a lot of the sky glow and often let a feeble galaxy punch through the haze and become detectable to the eye. (I have about 200 observations that would have been impossible without my Orion Sky Glow filter.) But beware—no filter will make objects *brighter* to see. There are special filters (like the O-III that actually enhance the detail you can see in gaseous emission nebulae, but they have little impact on most galaxies, which are composed not of narrow-band emission spectra but of broad-band stellar spectra. Also, the use of almost any filter will rob the optical system of almost one magnitude of light and cause false colors to be seen in stars. When you consider that the purpose of most broad band filters is to *enhance contrast*, one can usually afford a magnitude loss in galaxies if the sky behind them gets dark enough to let the weakened light punch through. For general sky glow, consider a broad band filter that cuts out the wavelengths associated with mercury vapor and high-pressure sodium lights. Metal halide lights are a different problem, as they emit a richer spectrum than the rather narrow bands of mercury vapor and high-pressure sodium lights. Low-pressure sodium lights are monochromatic and hence of no serious threat to sky glow. Narrow band filters, like the O-III or UHC, have special tasks behind their design, and although they can make the sky background velvety black, they can also filter out much of the light of stars and galaxies.

Rating the Sky Conditions

Related to sky glow is the sky's general transparency. For this purpose, I use a scale of 1 (cloud cover) to 5 (as clear as it gets from my observing location).

One must also contend with the turbulence of the air, a factor astronomers generally call "seeing." A large instrument with poor seeing is no better off than a small one with good seeing in many cases. In fact, larger telescopes often suffer more from bad "seeing" than smaller ones since there is more air in the column represented by the telescope's objective as the base of a cylinder of air that reaches from ground level to the top of the atmosphere. Within this column of air, there will be thousands of small eddy currents or pockets that each refract light a tiny amount. The combined effect of hundreds or thousands of such minute refractions is a boiling image that never holds steady. I generally rate seeing from 1 (dismal) to 5 (perfect).

Other Tricks of the Trade

While observing, use a filtered light for reading sky charts, making notes, sketches, and so on. The best color, of course, is red, as red lacks the quantum power to harm night vision. (However, I have also experimented with dimmable *green* light-emitting diodes (LEDs) and found no degradation in night vision with dim green light.) Personally,

I use a small variable output flashlight with two red LEDs and, for other purposes, two white LEDs. If you use a computer at scope side, be sure the software has a "night vision mode." *TheSky*, from Software Bisque of Golden, Colorado, is my program of choice, and it has a good night vision mode. (It is also the program that I used to generate all the star charts in this book. These charts were generated with the permission of Software Bisque.) As an added protection to my night vision, I also use a red Plexiglas cover for my laptop's screen. This added layer of filtering further reduces the intensity of light coming off the laptop's display.

Another visual acuity trick is to use a black cloth to cover your head and eyepiece region (similar to the old photographer's camera drape). I use a 3-foot square piece of black muslin for this and find it is a help when the moon is in the sky.

You will also enjoy greater light sensitivity if you do not fatigue your eyes. One of the greatest causes of fatigue is squinting the unused eye while you are at the eyepiece. I have learned over the years how to keep both eyes open while at the eyepiece and ignore what the unused eye is seeing. Until you can get to that level of visual discipline, feel free to use an eye patch (available at most pharmacies).

While at the eyepiece, there are a few things you can do to sharpen your eye's sensitivity as well. The retina is composed of rods and cones. Rods pick up light and contrast, while the cones detect color. Since our eyes are so color sensitive, the center of the retina is made mostly of cones. The rods lie on the periphery. If you want to see more at the eyepiece, learn how to look to the *side* of the target, not directly at it. Looking to the side (using what is called *averted vision*) lets the photons fall on the rod-rich part of the retina. Often, I have seen a galaxy with averted vision that disappears when I look directly at it.

The eye is also very sensitive to movement. I have found that at times when averted vision fails to detect a faint galaxy that moving the telescope slightly in declination or even tapping the tube to induce a little vibration will allow me to detect the object.

Still another way to heighten your visual acuity is to train your eye the way an artist trains his or her eye. Start drawing eggs. Keep drawing them until the drawing actually *looks* like an egg! When you reach that level of skill, your eye will see detail in galaxies, nebulae, and the like that you missed earlier.

Consider keeping a sketch book of your observations. Draw what you see (even a dense open cluster or rich globular). It does not have to be perfect or of museum quality. But you will find that drawing what you see forces your eye to concentrate on what is really there. I use good quality drawing paper (you can get artist's sketch pads at any art store) and a soft pencil and paper smudging stick. (The smudging stick helps to blur out details and results in a realistic rendering of galaxies and nebulae.)

In viewing the many extremely faint and tiny galaxies listed in this book, bear in mind that all that can be seen of many galaxies is the bright nucleus (especially when observing from light-polluted sites, like my suburban observatory). The arms of spiral galaxies are often so faint as to be undetectable in small telescopes. This means that for many galaxies, you will be searching for only the tiny nucleus. Such nuclei look like fuzzy stars in the eyepiece—they do not seem to focus quite as sharply as the stars in the field. Such "fuzzy stars" are what you will be looking for. When you have what you suspect to be a galactic nucleus in the field, run the magnification up to as high as the conditions will allow (I have used 700× before). If the object really is a galaxy, you will not be able to focus it no matter what you do. If it is a star, you will know soon enough.

Magnify, Magnify, Magnify

An old rule of thumb in astronomy is that the practical magnification you can use with a telescope is 30 times its aperture in inches (or 12 per cm). But how many people have the same size thumb?

I have learned through experience that when the sky is clear enough and transparent enough, I have run my C-8 up to 700× (88 times the aperture). I even ran the C-11 up to 1450× one night—some 132 times the aperture! Do not let someone tell you that 30× per inch is your limit. Push your scope as far as the seeing will allow, and you may be amazed at what you see!

Seeing Galaxies

Astronomers have developed numerous classification systems over the years to describe galaxies, ranging from Edwin Hubble's "tuning fork" diagram to the complex but comprehensive system developed by Gerard De Vaucouleurs. But to be quite honest with you, these systems are of high value only to those with really large apertures and the means to capture the precious light they collect.

To the average backyard amateur, galaxies come in three basic types: fuzzy stars, amorphous blobs of faint light, and a combination of the two.

Figures 3.1 through 3.3 illustrate these. Figure 3.1 shows the famous "fuzzy star" galaxy. These are galaxies that have bright and rather small nuclei (usually spirals or

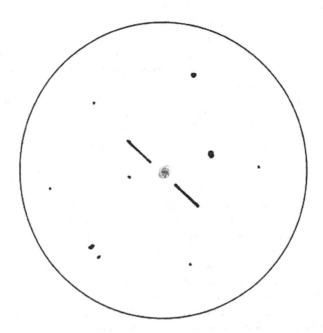

Figure 3.1. The infamous "fuzzy star" galaxy.

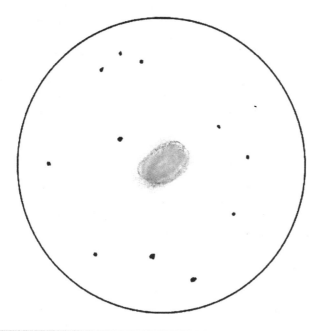

Figure 3.2. The "amorphous blob" galaxy.

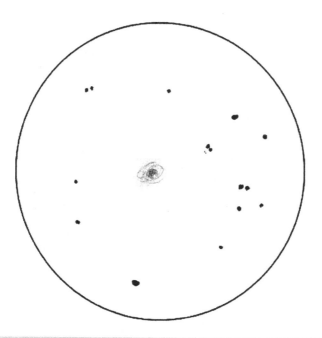

Figure 3.3. The "fuzzy star surrounded by a halo" galaxy.

barred spirals and Seyferts) and very faint or nonexistent arms (such as ellipticals or spheroidals). All you can normally detect of such galaxies from a suburban location with a moderate size instrument and good seeing is the bright nuclear region. In many of my descriptions, you will see "fuzzy star" noted. In most cases, you will need a good finder chart to confirm you have the *right* fuzzy star and not a faint field star that just looks fuzzy. (The charts on the CD are of high enough detail and quality to make this easy for you.) My usual method when I detect a fuzzy star galaxy is to switch to high powers (usually 230 or more) and see if I can get a sharp stellar object or a star that would not quite focus. If it focuses sharply, it is a star. If it still looks a little blurry while the other stars in the field are sharp, it is the nucleus of a galaxy (or an elliptical or spheroidal galaxy).

The type shown in Figure 3.1 is a fuzzy star type of galaxy. (To help you pick it out from the background stars, I have placed bracketing lines on either side of it.)

The type of galaxy shown in Figure 3.2 is the "amorphous blob of faint light" type of galaxy. Here, the galaxy has a much more subdued nucleus and/or brighter arms (often full of HII regions). This type is rather rare from suburban locations because if the skyglow is strong enough to drown the arms of a galaxy, it usually can drown an amorphous blob too.

The type shown in Figure 3.3 is a hybrid type, the fuzzy star surrounded by a halo. This is the normal pattern for spirals with bright arms, although it too is difficult to detect from suburbia.

Of course, a few galaxies—such as the brighter Messier galaxies—are strong enough to show considerable structure even in suburban conditions. I have, on rare occasions, detected spiral structure in M51 from my Kansas City back yard, and detected heavy mottling in many of the galaxies in Leo and Virgo, but such nights are not very common for me in the Midwest United States.

One very important conclusion to draw from this discussion is that whereas a catalog may say that a particular galaxy has certain dimensions, you must remember that these are the dimensions as measured on a photographic plate or CCD image. Such plates are far more sensitive to faint light than the eye, so do not be surprised if while observing a galaxy from a suburban location, you can only detect a blur of light that is $1/4$ or less the size the catalog suggests. You may, after all, only be pulling out the nuclear region.

One way to help gain back *some* of the detail the light in your telescope wants to give you is to observe with patience. Your eye cannot, like film or a CCD chip, store photons and build an integrated image. It is "live" light. But you can often start to pick up more and more detail in faint galaxies if you sit patiently at the eyepiece and wait for the sky to come to you. The sky usually has transient moments of brilliant clarity in the midst of mundane murkiness even on the worst of nights, and if you are patient, you will get brief glimpses of a galaxy with stunning detail before it fades back to a boiling pot of cauliflower.

CHAPTER FOUR

Double Stars Galore!

Who Was that Masked Double Star?

Here is a technique that will help you split difficult double stars and see greater detail in planets. A difficult double star can be one in which there is a great difference between the brightness of the main star (primary) and its companion, or where two bright stars are very close together. The glare of a bright primary can often drown a feeble companion. Also, when two stars are very close together, the image of the fainter star may actually lie on one of the diffraction rings of the primary.

This is where an *objective mask* can help. An objective mask induces interference in the light path of the telescope and can turn the Airy disc and diffraction rings of a normal view into a sharp Airy disc with no rings. In my case, I built a *hexagonal* mask out of foam core board. Placing it over the business end of my C-8 results in stellar images that are sharp Airy discs with six spikes radiating outward from the Airy disc (Figure 4.1). If a faint companion lies on the primary's diffraction rings, the mask often removes the diffracting rings and lets the faint companion pop into view. If the spike sits on the companion, rotating the mask a few degrees often brings the companion into view.

This technique is mostly of value to centrally obstructed telescopes, such as Newtonians and SCTs. Experimentation by colleagues who have refractors shows that the mask is of little value to such instruments.

Figure 4.1. Objective mask on the C-8.

It was by the use of such a mask that I detected the elusive companion of Sirius and the notoriously difficult *blue*[1] companion of Antares.

Double Jeopardy

Observing double and multiple stars can be at the same time one of the easiest and yet most challenging activities for an amateur astronomer. It can also be a very rewarding area of study from an aesthetic sense. If you have the instrumentation for it and the required skill, you can even make measurements that can contribute to orbital solutions that help us refine our models of stellar evolution.

Double and multiple stars consist of a main star (called the primary, or in the case of multiples, A) and a companion (sometimes called *comes*, or B) or companions (*comites*, or B, C, D, etc.). Double and multiple stars are stars that are truly gravitationally bound and traveling through space together. Sometimes, two stars will happen to line up along the same line of sight and *appear* to be a double star system. However, they are not gravitationally bound at all and represent a chance alignment. Such a pair is called an *optical double* to distinguish it from a true double, which is often also called a *binary system*. I have made every attempt to point out which double stars in my list are (or may be) opticals.

Most binaries have members that are visible in a telescope, but a large number of what appear to be single stars are actually very close binaries. They are so close together in space that they appear as one star from earth. The fact that they are two stars instead of one is revealed by the spectroscope as the emission lines of both stars show up in

---

[1] Some observers claim that the companion is green, but there are no green stars. The green hue described by some is probably a "contrast effect" of the blue being set against the deep orange-red of Antares.

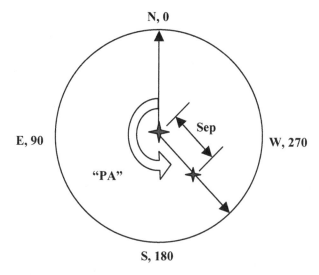

Figure 4.2. Position angle and separation in double stars.

the combined spectrum and these lines shift left and right as the stars approach or recede from earth as they orbit one another (due to the Doppler shift). These ultra-close pairs, not resolvable in any amateur telescope (and only a few of which can be resolved by huge instruments), are called *spectroscopic binaries*. In a few rare cases, these pairs have been resolved using a technique called *speckle interferometry*. A small list of others have been "resolved" by lunar occultations as the light output of the system drops in steps rather than the dramatic "on/off" of a single star occultation. These stars are often called *occultation binaries*.

In this book, you will be observing normal binaries and optical pairs. For these stars, there are two measurements that are crucial: *position angle* (PA) and *separation*. PA refers to the angle between the A and B stars measured counterclockwise from north, using the A star as the center of the frame of reference (Figure 4.2). In this method, north is 0°, east is 90°, south is 180°, and west is 270°. So a PA of 225 indicates that the star in question is southwest of A. But before you can know what to expect at the eyepiece, you need to know how your telescope presents the image.

With both of my SCTs, I use a right-angle "star diagonal," which produces an upright but mirror image of the star field. So for my purposes, the PA scale runs clockwise from the top of the field, making east the right side of my field, west the left, north at the top, and south at the bottom. Other telescopes, however, produce inverted and mirror images, or inverted but normal images, and so on. The easiest way to find out how your scope presents the field is to center a bright star in the field of view and turn the declination slow motion control back and forth and note which way the star moves. If you turn your declination knob so as to move the telescope north and the star drops to the bottom of the field, your scope does not invert the image; if it rises, it does invert the image. For the mirror imaging check, recenter the star and turn off the clock drive (if you have one) and let the star begin to drift. It will drift toward the west, so whichever side of the field the star drifts to is west. If the scope makes a mirror image (like mine), the star will drift to the left; if it does not make a mirror image, the star will drift to the right.

Separation is the angular distance between the stars. A separation of 30 seconds (30″) would mean that the two stars are half a minute (1/2′) (or 1/120th of a degree) apart. Combining separation with PA gives an accurate idea of what you can expect when you look into the eyepiece. For instance, if the components of a double star are listed as being 8″ apart in PA 40, you should look for a close pair with the B star lying about midway between north and east.

How Large Is the Field of View?

To fully appreciate the separation measurement, you need to know how large a field of view each of your eyepieces gives. There is an easy way to do this. You will need a stopwatch. Put one of your eyepieces in the telescope and center a fairly bright star in the field of view. (Most texts on this method suggest a star near the equator, but I prefer to use stars around 45° to 75° north declination for reasons you will soon learn.) Next, unlock the Right Ascension (RA) brake and use the slow motion controls to move the star so that it is just off the field of view on the east side of the field and lock the brake. Turn off the clock drive. As soon as the star creeps into view, start the stopwatch. Watch the star as it traverses the field, and when it exits the west side of the field, stop the stopwatch. Divide the transit time in seconds by 4 and multiply that result by the cosine of the declination to get the field width in minutes of arc. Do this about six times for each eyepiece and average the results. You will find a form for this calculation on the CD-ROM that comes with the book.

Example. You use δ Orionis (the west-most star of Orion's "belt") to test your eyepieces. With the first eyepiece, you get an average transit time of 180 seconds. Dividing 180 by 4 gives 45 minutes of arc. Since δ Ori's declination is −0018, the cosine is 0.9999863, so for all practical purposes, the field is 45′ in diameter. Suppose that for your next eyepiece, your average transit time is 118 seconds. The field of view would then be 29.5 minutes of arc. A third eyepiece, with an average transit time of 54 seconds, would have a field of view of 13.5′.

Had you used ε Persei (declination of +4000) for the transits, the transit times would have been longer—235 seconds, 154 seconds, and 70 seconds, respectively. The cosine of 40° is 0.7660444. Dividing the transit time for the first eyepiece by 4 produces 58.75, which when multiplied by the cosine of 40°, is 45′, the same result we obtained using δ Ori.

The use of stars farther north means that the transit will take longer to do. The advantage to this is that any errors in starting and stopping the stopwatch will be spread over a much longer increment of time, so the errors in timing will be smaller as a fraction of the total increment being measured.

If you plan on doing double star measurements, you need to add a correction factor to the process previously mentioned because double star measurement is one of the most precise forms of measurement known. You must correct for the difference between the earth's rate of rotation and the sidereal rate. To do this, multiply the field diameter you get by this method by 1.002739.

Starlight, Star Bright, What Color Are You?

The other details necessary for double star observations are their magnitudes and spectra or colors. The magnitudes are straightforward enough (although you should be aware that some of the double star discoverers of the nineteenth century were notorious for mis-stating the magnitudes of their stars).[2] The main difficulty you will encounter with magnitudes is where two stars of great magnitude difference are close to each other, in which case the objective mask I described earlier will be a useful tool along with very high magnification. And in some cases, a primary or companion could actually be a variable and the original discoverer did not know this. Therefore, the original magnitude estimate could be off by many magnitudes.

The color issue is more subtle. All stars have color, but most of the hues are so subtle that to the casual observer, they all look white. A star's color is a function of its surface temperature, which in turn is tied to its spectral class. The Morgan-Keenan-Kellman spectral class system (the pioneering work having been done by Annie Jump Cannon of Harvard College Observatory) runs OBAFGKM (often remembered by the mnemonic, "Oh, be a fine girl: kiss me"). Recent discoveries have added a few more classes (W, which has been dropped, and RNSLT).[3] The temperatures run in the same sequence. O stars are the hottest of all (surface temperatures of 70,000 K or hotter) and have a definite bluish to violet tint. B stars are a little cooler and also appear bluish. The A stars are cooler yet and appear bluish-white. F stars are cooler, and look white to most observers. G stars (like the Sun) are yellowish. K stars look orange, while M stars are red. But remember that these colors are subtle, and differences in eyes and atmospheric conditions can alter the colors any one observer might perceive. When I say that I saw a pair of stars as blue and orange, that is my assessment of the colors, based on subtle shades of color; the colors you perceive may very well be different.

You should also be aware that the eye tends to see fainter and fainter stars as more and more bluish in tint. This is a peculiar effect of the eye (the Purkinje effect), and not a true case of faint stars being blue.

Spectral subclasses run 0 to 9, where 0 is at the "hot" end of the class and 9 at the "cool" end. Thus a G0 star is just a little cooler than an F9. Other sub-codes include "comp" for composite spectrum (usually due to an unresolved binary), "e" for emission lines, "m" for metallic, "n" for broad lines (usually caused by rapid rotation), "p" for peculiar, "s" for sharp lines, "shell" for shell star (main sequence star embedded in a gaseous shell), "Si" for strong silicon lines (or other metals, using the chemical symbol for those metals), and "v" for variable.

---

[2] For example, The Reverend T. W. Webb, in his classic handbook *Celestial Objects for Common Telescopes*, Volume 2, presents a chart showing how the magnitude scales used by Smyth (a nineteenth century double star fanatic) compared to Friedrich Struve (the dean of double star discoverers), John Herschel, and Argelander. As an example of the wild variation in those days, a Smyth 10.0 magnitude is equivalent to a Struve 9.3, a Herschel 10.4, and an Argelander 9.4!

[3] When WRNS was added to the MKK taxonomy, the mnemonic was changed to, "Oh, be a fine girl: kiss me. Well, right now, sweetheart!" With the dropping of W and the addition of L and T, one can only imagine how someone will modify the mnemonic!

In the color taxonomy I use for this book, colors are represented by letters. The taxonomy is as follows:

R red

O orange

Y yellow

G green (rare in star colors; some including this author argue it does not exist at all)

W white

B blue

V violet

L lilac

D gold

In addition, the code can be modified by preceding the capital color letter with a lowercase modifier. A "p" should be read as "pale," while "d" is read "deep." Thus, pB is "pale blue" and dR is "deep red." In addition, using a lowercase color letter before an uppercase color letter indicates a blend of colors, as rO would be read as "reddish-orange."

Intense colors will be indicated by the addition of an exclamation point at the end of the code and very intense colors by two exclamation points. Thus R! would suggest "very strong red," while D!! indicates "extremely deep gold." A question mark indicates that the color is at best questionable. B? would be read as "blue, maybe." If all you see in the color field is a question mark, it means the star was too faint to get a reliable color estimate. Finally, if I could not detect the star, you will see "No" in the color field.

Putting It All Together

The double star STF 541 (STF being the code for Friedrich Struve—this star is number 541 in his Dorpat Catalog of 1827) lies in Canis Major and consists of a K0 primary accompanied by an F class secondary 23″ away in PA 44, with magnitudes of 8.0 and 9.0. What should you expect to see in the eyepiece?

You should see a moderately bright orange star and, 23″ away in the northeast direction, a fainter white companion.

Whose Double Is it, Anyway?

The double stars in this book will be listed by the name (or code) of their discoverer and the number which that pair occupies on his or her list. Whenever possible, the earliest known discoverer will be given the honors of top billing, with subsequent discoverers (or those who found later companions) listed under the "other names" field. When available, I will also list the star's position in Robert Aitken's master list of ADS numbers. The code for discoverers is as follows:

AG   Astronomische Gesellschaft Katalog of 1875.

A   Robert Aitken, 1864–1951, American astronomer and astronomer at Lick Observatory. Aitken brought order to the nonsystematic observations of Sherburne Burnham and others by using a rigorous program at Lick to confirm and/or discover 17,000 binaries down to magnitude 9.0. (This catalog was published in 1932.) After completing his massive catalog, he spent many years computing binary orbits.

Abetti   Giorgio Abetti, 1882–1982.

AC   Alvan Clark, 1804–1887.

Alden   H. L. Alden.

Algiers Obs   Algiers Observatory.

Ali   Alaaeldin Ali, assistant professor of astronomy at the United Arab Emirates University.

Aller   Ramon Maria Aller, 187–1966, Spanish priest and astronomer with an outstanding humanitarian biography.

Anderson   John August Anderson, 1876–1959, pioneer in using interferometry at Mt. Wilson Observatory for measuring double stars.

Aravamudan   Shri Aravamudan.

Argelander   Friedrich Wilhelm August Argelander, 1799–1875, Prussian astronomer and assistant to Bessel. His extensive measurement of star positions resulted in the 1863 *Bonner Durchmusterung*, still a mainstay of modern astronometric research.

Bur   Sherburne W. Burnham, 1838–1921, one of the most prolific and keen-eyed binary star discoverers. S. W. Burnham was not a professional astronomer but was highly esteemed by the professional community nonetheless. He became interested in astronomy as a Confederate soldier in New Orleans during the American Civil War, but continued his career as a reporter and clerk. After the Civil War, he settled in Chicago, only a few hundred yards from the Dearborn Observatory. In 1869, he met Alvan Clark, from whom he ordered the 6-inch refractor with which he would discover hundreds of new double stars. He found his first new pair, B40, on April 27, 1870. When he had a list of 81 new pairs, he sent the list off to the Monthly Notices of the Royal Astronomical Society. This was followed by two more lists a year later. For these first lists, Burnham had no filar micrometer, so the measurements were made by the Italian astronomer Dembowski. While on a vacation in 1874, he got to use a 9.4-inch telescope at Dartmouth College Observatory and even got to use the 26-inch refractor at Washington one night. During this 10-day period, he added 14 stars to his list. From that time forward, he was able to use many of the larger American telescopes, including the 36-inch Lick and 40-inch Yerkes instruments. From 1888 to 1892, he was officially on the staff at Lick, although he kept his day job and would commute to Williams Bay, Wisconsin on weekends to use the Yerkes instruments. During his career, he discovered 1274 new doubles, many of which are true binaries. In 1894, he received the gold medal of the Royal Astronomical Society for his work, and in 1904, the Lalande Prize from the Paris Academy of Sciences, as well as honorary degrees from Yale and Northwestern Universities. Not bad for an amateur!

Baillaud   Rene Baillaud, 1885–1977, French astronomer.

Baize　Paul Baize, 1901–1995. Trained as a medical doctor and receiving his degree in 1924, Paul has practiced most of his life as a pediatrician. His astronomical work, of prodigious quantity, has been purely amateur. He began double star work in 1925 with a 10.8-cm refractor with a micrometer he built himself. Between 1925 and 1932, he made 3834 measurements that were published in *Les Journal des Observateurs*. Starting in 1933, he used the 30.5-cm equatorial refractor in the West Tower of the Paris Observatory and completed 11,332 measurements by 1949. From 1949 to 1971, he used the 38-cm telescope in the East Tower and produced an additional 8878 measurements. He has also calculated the orbits of some 200 binaries and published numerous articles for astronomical journals.

Ball　R. S. Ball, 1840–1913.

Barnard　E. E. Barnard, 1857–1923.

Barton　S. G. Barton.

Bemporad　A. Bemporad, 1875–1945.

Bergh　Sidney van den Bergh, born in 1929, Canadian astronomer.

Bhaskaran　T. P. Bhaskaran, born in 1889, Indian astronomer.

Bigourdan　Guillaume Bigourdan, 1851–1932, French astronomer.

Bird　F. Bird.

Bos　Willem H. van den Bos, 1896–1974. Born in Rotterdam on September 25, van den Bos was attracted to astronomy and double stars at an early age. He started studying astronomy at Leiden in 1913 and was taught by such men as Hertzsprung and deSitter. He earned his PhD in 1925 (physics/mathematics) and traveled that year to South Africa to serve as a guest observer at the Union (Republic) Observatory of Johannesburg. In 1930, he joined the staff at Union Observatory and remained there until his retirement in 1956. He continued double star research on a private basis until ill health forced him to stop in 1966. By that time, he had made 74,000 visual measurements, discovered some 2900 new pairs, and computed 150 orbits. Known for great speed (he could make 20 measures per hour), he was also a very keen observer.

Bottger　G. van Bottger.

Bowyer　W. M. Bowyer.

Bpm　Burnham's proper motion pairs.

Brisbane Obs　Brisbane (Australia) Observatory.

Burton　Charles Edward Burton, 1846–1882, tutored in the art of mirror making by none other than Lord Rosse. He went on to produce fine glass mirrors up to 15 inches in diameter.

Chevalier　P. S. Chevalier.

Cogshall　W. A. Cogshall, 1900–1944, professor of astronomy at Indiana University.

Cordoba　Cordoba Observatory (Argentina).

Courtot　Jean Francois Courtot.

Couteau　Paul Couteau, born in 1923. A Frenchman, Paul Couteau knew he wanted to be an astronomer by age 11. In 1949, he earned his PhD from the Astrophysical Institute of Paris (thesis on white dwarfs), and began his observing career professionally at Nice (1951–1967), making some 12,000 measurements. He also spent time at Yerkes Observatory (5 months in 1961). Since 1965, he has systematically measured the stars of the BD Catalog zone +17 to +53 (some 170,000 stars are in this band). He had observed 102,000 stars by 1984 and discovered 2200 pairs, half of which are closer than 0.5″ of arc.

CPO Cape Observatory (South Africa).

Dawes William Rutter Dawes, 1799–1868. A doctor and clergyman, Dawes was always in poor health. The bulk of his work is contained in the *Catalog of Micrometrical Measures of Double Stars*, which is part of volume 35 of the Transactions of the Astronomical Society of London.

Dawson B. H. Dawson.

Dembowski Baron Ercole Dembowski, 1812–1881. Born in Milan on January 12, he was of noble ancestry on his father's side. His early life was in the military and only turned to astronomy after meeting an astronomer (Antonio Nobile) sometime after 1843. He published his first list of pairs in 1851, which consisted of accurate measurements of 127 of F. G. Struve's list. Beginning in 1858, he made a thorough update to Struve's Dorpat Catalog.

Doberck W. A. Doberck.

Donner H. F. Donner.

Doolittle Eric Doolittle, 1869–1920, an American. Eric graduated from Lehigh University in 1891 as a civil engineer and became an instructor of mathematics at Lehigh and later Iowa University. From 1896 to the end of his life, Eric worked at the University of Pennsylvania's Flower Observatory where he used the new 18-inch Brashear refractor to create four large volumes of measurements.

Dorpat Obs Dorpat Observatory.

Duner N. C. Duner.

Dunlop James Dunlop, 1795–1848, Australian astronomer and pioneer in charting the southern skies.

Egbert H. V. Egbert.

Engelmann R. Engelmann.

Espin Thomas Henry Espinall Compton Espin, 1858–1934. He was a cleric and Vicar of Town Law in County Durham, England. He became interested in astronomy after 1874. His first work was a catalog of 3800 red stars (1885–1899). In 1900, he started double star work with the 17-inch Calver reflector, finding 2575 pairs, most of them wide ones. He also discovered Nova Lacertae on December 30, 1910.

F Brown F. Brown.

Fender F. G. Fender.

Filipov M. L. Filipov.

Finsen William Steven Finsen, 1905–1979. An Icelander (though born in Johannesburg) and a nephew of a Nobel Prize winner, Finsen assisted van den Bos at the Union Observatory. Later, using instrumentation of his own design, he made 13,000 measurements of 8117 stars between −75 and +20 degrees declination. From this work, 73 new pairs were found, 11 of which have orbital periods of less than 21 years. He also made 6000 measures that were too close for the micrometer.

Fox Philip Fox, 1878–1944.

Franks W. S. Franks, 1851–1935.

G Struve George Struve.

Gallo J. Gallo, 1882–1965.

GAn George Anderson.

Gauchet P. L. Gauchet.

Giacobini M. Giacobini.

Giclas Henry Lee Giclas, born in 1910. American astronomer who did duties at Lowell Observatory.

Gilliss James Melville Gilliss, self-taught US Naval lieutenant who worked at the US Naval Observatory.

Glaisher James Lee Glaisher, 1848–1928, English mathematician and astronomer.

Glasenapp Sergei Pavlovich Glasenapp, 1848–1937. Born Glazenap, Sergei was a Russian astronomer and early pioneer in the accurate measurement of the speed of light.

Goyal A. N. Goyal.

H William Herschel, 1739–1822. Herschel's first double star work was to find good candidates for parallax measurements. His first list had 269 stars on it, only 42 of which had been known previously. His painstaking measurements of these stars convinced him that the changes he saw were not due to parallax after all, but to tiny changes as the stars seemed to orbit one another (a radical concept at the time). Herschel subclasses include I (difficult), II (close but measurable), III (5″ to 15″), IV (15″ to 30″), V (30″ to 60″), VI (1′ to 2′), and N (the 1821 catalog).

h John Herschel, 1792–1871. John's early work was to remeasure his father's lists to confirm the elder Herschel's binary star theory. Later, he teamed up with James South, and together they published some 380 new pairs. John's expedition to the Cape of Good Hope (1834–1838) was a milestone for astronomy, the younger Herschel cataloging some 1202 new pairs.

Harvard Harvard Observatory.

Hastings C. S. Hastings, 1848–1932.

Haupt H. Haupt.

H Wilson H. C. Wilson, American astronomer, editor of *Popular Astronomy*, professor of astronomy at Carleton College.

Heintz Wulff Heintz, professor Emeritus of Astronomy at Swarthmore College, work included directorship of Sproul Observatory. One of the leading solvers of binary star orbits.

Hernandez S. Hernandez.

Hertzsprung Ejnar Hertzsprung, 1873–1967, Danish astronomer and codeveloper of the Hertzsprung-Russel (H-R) diagram.

Hh An obscure index catalog produced by William Herschel.

Hipparcos Hipparcos space astronometry mission (ESA, 1992).

Holden Frank Holden, 1917–1992.

Holmes E. Holmes, born in 1919.

Hough G. W. Hough, 1836–1909.

Howe   H. A. Howe, 1858–1926.

H Struve   H. Struve.

Hussey   William Joseph Hussey, 1862–1926. Originally a professor of mathematics at the University of Michigan, Hussey went to Palo Alto, CA, where he was a professor of astronomy at the Leland Stanford Junior University. While there, he made frequent visits to Lick Observatory, joining the staff in 1896. Working with Robert Aitken, he discovered 1327 new pairs in 6 years. In 1905, he returned to the University of Michigan and, in 1911, became director of the LaPlata Observatory in Argentina. His great ambition was to see a great telescope for southern hemisphere double star observations to be built.

Innes   Robert Thorburn Ayton Innes, 1861–1933. A Scotsman by birth and education, all of Innes's amateur astronomical career was in the southern hemisphere. His capstone work, the *Southern Double Star Catalog*, was published in 1927, and it contains 1613 of his discoveries plus thousands of other pairs.

Jacob   W. S. Jacob.

Jessup   M. K. Jessup.

Jonckheere   Robert Jonckheere, 1889–1974. Robert began his career as a double star observer in 1905. His first published list (1908) contained 40 pairs. In 1962, he published his major work, *General Catalog*, which contains 3355 new pairs he discovered since 1906.

Kazeza   S. M. Kazeza, member of the Hipparcos Input Catalog team.

Knott   G. Knott.

A Krueger   A. Krueger.

Kruger   E. C. Kruger.

Kuiper   Gerard P. Kuiper, 1905–1973.

Kustner   Friedrich Kustner, 1856–1936, German astronomer.

Lamont   Johann von Lamont, 1805–1879, German astronomer, born in Scotland, director of Bogenhausen Observatory and professor of astronomy at Munich University, best known for his magnetic surveys and maps of Europe. He observed Neptune in 1845 and twice in 1846, but did not realize it was a planet!

LaPlata   La Plata Observatory, Argentina.

Lalande   Joesph Jerome Lefrancois de Lalande, 1732–1807, French astronomer.

Larink   Johannes Larink.

Lau   H. E. Lau, 1879–1918.

Leavenworth   F. P. Leavenworth, 1858–1928.

Leonard   F. C. Leonard, 1896–1960.

Lewis   Thomas Lewis, 1856–1927. Lewis joined the staff of the Royal Observatory in 1881. His major achievement is the book *Measures of the Double Stars contained in the Mensurae Micrometricae of F. G. W. Struve* (1906).

Lhose   O. Lohse.

LDS   W. J. Luyten, 1899–1994.

Madler   J. Madler.

Mason   H. C. Mason, 1813–1936.

Milburn   W. Milburn.

Miller   J. A. Miller, 1859–1946.

Mitchell   S. A. Mitchell.

Morgan   H. R. Morgan.

Motherwell   R. M. Motherwell.

P Muller   Paul Muller, born on 1910. Paul joined the staff of the Observatory of Strassbourg in 1931. He has measured pairs with an instrument of his own design, the double image micrometer. He has made over 11,000 measurements and calculated 90 orbits.

Olivier   C. P. Olivier, –1975.

Opik   E. Opik.

OS   Otto Struve, 1819–1905. At 18 years of age, Otto became his father's assistant at Dorpat. His double star work includes the discovery of 547 pairs. The OS stars are his original Pulkovo catalog of 1843.

OSS   Otto Struve Index Catalog, 1843, containing wide pairs.

Osvalds   V. Osvalds.

Paloque   E. Paloque.

Panjaitan: Edward Panjaitan

Pannuzzio: R. Pannuzzio, Italian astronomer

Perrine: Charles Dillon Perrine, 1867–1951. Argentine-American astronomer, discover of Himalia and Elara, Moons of Jupiter.

Perth Obs   Perth Observatory (Australia).

Pettitt   Edison Pettit, 1889–1962.

Pocock   R. J. Pocock, 1889–1918, Indian astronomer.

Popovic   C. Popovic, 1910–1977.

Pourteau   Abel Pourteau, French astronomer.

Pritchett   H. S. Pritchett.

Pulkovo   Pulkovo Observatory.

Przbylloc   E. Przbylloc.

Rabe   W. Rabe.

Roe   E. D. Roe, Jr, born on 1929.

Rossiter   Richard Alfred Rossiter, 1886–1977. The dean of southern hemisphere double star astronomy, Richard was born in New York but moved to South Africa in 1926. (During the trip, William Hussey, who was accompanying Rossiter, died suddenly in London.) His work produced a list of 7368 southern pairs, of which 5534 were his own discoveries, still a record.

Rousseau: J. M. Rousseau.

Russell   H. C. Russell, 1836–1907.

STF   Freidrich Wilhelm Struve, 1793–1864. The STF stars are his Dorpat Catalog of 1827. The "rej" that appears on some of the stars (as it does also in the OS stars) means that Freidrich rejected the pair from his original catalog because he thought it was too far apart to merit further study. In a 2-year project, Struve made 10,448 measurements and produced a list of 3112 pairs of which 2343 were new.

Scardia   M. Scardia.

Scheiner   J. Scheiner, 1858–1913.

Schjellerup   H. C. F. C. Schjellerup.

Secchi   Father Angelo Secchi, 1818–1878. Best known for his work in stellar spectra, Secchi also made contributions to double star work, producing a catalog of 1321 doubles in 1860.

See   Thomas J. J. See, 1866–1962.

Sh   James South, John Herschel joint catalog of 1824.

SI   F. W. Struve's First Index Catalog.

SII   F. W. Struve's Second Index Catalog.

Skinner   A. N. Skinner, –1918.

Smart   W. M. Smart, –1975.

Sola   J. Comas Sola.

Soulie   G. Soulie.

South   James South, 1785–1867. South was a surgeon who later became interested in astronomy. Marrying into great wealth made the pursuit of his hobby an amateur's dream.

Stein   J. Stein, 1871–1951.

Stone   Ormond Stone, –1933.

Swift   Lewis Swift, 1820–1913, American astronomer; working at the Warner Observatory (16-inch reflector), he discovered 600 nebulae.

Tarrant   K. J. Tarrant.

Torino Obs   Torino Observatory.

Tucker

Tycho   Tycho astrometric mission (ESA, 1992).

Upton   W. Upton.

USNO   US Naval Observatory, Washington, DC.

Van Biesbroeck   George van Biesbroeck, 1880–1974. One of the greatest double star observers of all time, he was active for over 70 years. George is best known for his painstaking measurements and rechecking of earlier catalogs for orbital changes, not so much for discovering new pairs. (He made 35,915 measures during his life.)

van de Kamp   Peter van de Kamp.

Vatican Obs   Vatican Observatory.

Vilkki   E. Vilkki.

Voute   John George Erardus Voute, 1879–1963. Voute began his astronomical career at Leiden Observatory in 1908, publishing a double star orbit in his first year. He spent most of his career at the Cape Observatory. He made 26,126 measures over his lifetime.

Walker   R. L. Walker.

Ward   I. W. Ward.

Webb   T. W. Webb, 1807–1885.

Weisse   M. Weisse.

Winnecke   F. A. Winnecke.

Wirtz   Carl Wirtz, −1939.

Worley   Carl Worley. −1997.

Can this Magnitude Be Right?

One thing you will notice right away when observing double stars is that in many cases, the magnitudes listed in the catalogs just do not look right! And they are not!

So what is going on? In most cases, you will find that astronomers who worked in the late 1800s often seemed to badly mis-state the magnitudes of their stars. For example, John Herschel was notorious for this, as were such observers as Gillis, Baillaud, Scheiner, Stein, and Chevalier, just to name a few. Were these people trying to publish lists of double stars in such a way as to make their pairs *look* brighter than they actually were? Were there "bragging rights" for magnitude in the 1800s?

No, there were just inadequate photometric catalogs and inferior instruments. For instance, some of John Herschel's work was done with speculum mirrors. Speculum is an alloy of tin and copper, and although it is fairly reflective, it does tarnish quickly and require frequent polishing. Even at its most brilliant, though, it still performs far more poorly than even mediocre glass. So a star that was in reality perhaps 10th magnitude would appear much dimmer than that to John Herschel, and he may have overcompensated for this when listing magnitudes. Indeed, some of his fainter pairs, showing a 16th magnitude companion, have been checked with modern photometric equipment. In some cases, John stated the magnitudes two or three digits too high. Sometimes, a Herschel "10" is really closer to a 12.00 in the modern Pogson scale, whereas a "16" may be closer to a 13.00 and so on.

Second, in the 1800s, there were not a lot of high-quality photometric catalogs for astronomers to work with. If an astronomer found a new double star in a faint field, he may not have any stars on the field chart that were of known magnitude. So he was on his own when it came to estimating magnitude. If you have ever tried to estimate the magnitude of a variable star with a good field chart showing the magnitudes of nearby stars, you can only imagine how hard it would be to accurately assess the magnitudes of a pair with *no* reference stars nearby!

Third, either the primary or the companion could be a variable star, and this fact may not have been known to the original discoverer. So on the night of discovery, a companion may have appeared as 11.6 magnitude when it was in fact a variable and in the declining part of its light curve. If you or I were to view this star tonight, we might catch it at a different part of its light curve and see it as 12.6 magnitude, or 10.2, and so on.

Double Stars Move Over Time!

Double stars are dynamic systems, i.e., they move over time even as the earth and other planets revolve around the Sun. Thus, the characteristics of a double star at one time may be different from a later time. I try to show this by two cues in this book. One is the "Year" field. This information tells you the year of the double star measurement I use in this book. Unless otherwise stated, all data is from the 2001 edition of the *Washington Double Star Catalog* (or WDS), the "Bible" of double star research.

If there is only one measurement on record, you will see the separation and position angle data without modifiers.

But if there is more than one measurement, I will usually tell the first measurement (in the Notes area) and the most recent measurement in the double star's data table. If the latest measurement is the same as the first one, I will follow the data with an equal sign (=). If the later data is greater than the first data, a plus sign (+) will be used, while;f the later data is less than the first data, a minus sign (−) will be used.

Perhaps a couple of examples will help. Suppose that the measures for a particular double star in 1844 were 36.2″ of separation at a position angle of 121 degrees. Suppose also that the WDS shows a 1994 measure of 36.2″ at 121 degrees. In this case, you would simply see 36.2 = in the separation field and 121 = in the position angle field.

But if the WDS had shown 1994 measures of 38.3″ at 126 degrees, you would see 38.3 + and 126 + in the separation and position angle fields respectively.

And if the WDS had shown 1994 measures of 38.3″ at 112 degrees, you would see 38.3 + and 112 −.

If the measures have changed significantly from the first ones, a single exclamation point (!) will be used. If the measures have changed greatly, two points will be used (!!).

Finally, be careful that you do not make the assumption that a + sign means a measure is currently increasing (or a − sign indicates that it is decreasing). The + and − signs are only relative to the first measure. If a pair is a true binary and it has recently passed apastron, there is a very good chance that recent measures will be greater than the original ones, yet still smaller than the measures when the pair was at maximum separation.

The following "clips" from typical double star listings show what I mean.

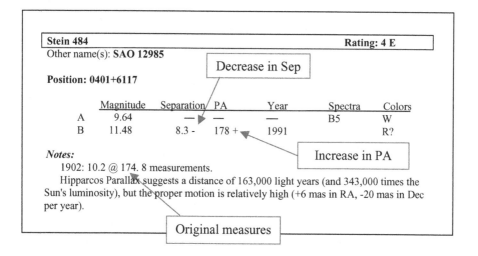

Holmes 3				Rating: 3 E		

Other name(s): **ADS 2984**

No change in measurement

Position: **0408+6220**

	Magnitude	Separation	PA	Year	Spectra	Colors
A	9.39	—	—	—	B0	W
B	10.69	5.4 -	215 =	1925		W?
C	12.40	14.0	129	1902		?

Notes:
   1902: 5.6 @215. 4 measurements.

Year of most recent measure

When referring to directions in the observing notes, I will use abbreviations for the cardinal directions; therefore, E is east, W is west, and so on. I will also use a lower case "m" attached to a number to represent "magnitude." Thus, 8m would be read as "8th magnitude."

CHAPTER FIVE

Deep-Sky Objects

Peering Way Back in Time

This book is not about double stars and double stars only (although they make up the bulk of the work). There are plenty of other objects in the sky to draw the attention of an amateur observer!

The so-called "deep-sky objects" (as if double stars are not deep enough in the sky!) consist of nebulae, star clusters (both open and globular), planetary nebulae, supernova remnants, and galaxies.

Your telescope—no matter what its size—is also a "time machine." As you look at the wonders in the night sky, the light you are seeing at this very moment began its journey at some time in the past. The farther away the object is, the older the light that strikes your eye that night. To me, one of the great thrills of amateur astronomy is the realization that as I am observing a faint galaxy's nucleus, a chemical reaction is taking place on my retina, started by the photons from that galaxy striking it and releasing a complex cascade of enzymes and other biochemicals that ultimately results in my brain "seeing" something. To think that at this exact moment my brain is responding to light that, in some cases, is older than the dinosaurs is a concept that often leaves me at a loss for words!

There are several fields of information supplied for every deep-sky object in this book. First, of course, is the *magnitude* of the object. But be careful—this can be deceptive. The magnitude for an extended deep-sky object is its "point source" magnitude. In other words, if all the light emitted by the object came from a single star-sized point, how bright would that point be? Many times I have searched for a 9th magnitude star cluster to find that I was dealing with a small but very faint grouping of stars, none of which were 9th magnitude themselves. In fact, if you wanted to get a rough idea of what to expect on some of the nebulous or galactic objects, find a star that is the magnitude of the object and defocus it until its bloated image is the size of the deep-sky object. That is what you will be looking for! (Try bloating a 10th-magnitude star into a 30-inch diameter blob to simulate a faint elliptical galaxy. It is not an easy thing to see!)

Related to this is the concept of *surface brightness*. This is a more reliable indicator of how difficult it will be to see the object, but surface brightness is only useful for nebulae and galaxies, as it indicates the approximate magnitude of any point on the surface of the object.

Here is an example of a deep-sky listing that shows the listed magnitude and surface brightness.

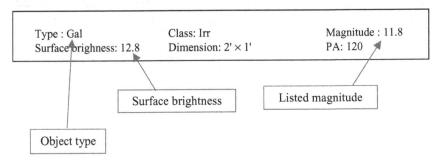

The *types of deep-sky objects* follow this code:

Gal   galaxy
Pn   planetary nebula
Gc   globular cluster
Oc   open cluster
Gn   bright galactic nebula
Dn   dark nebula
QSO   quasi-stellar object
SNR   supernova remnant

The *size* of a deep-sky object will be shown in minutes of arc. (In some cases, the dimensions will be shown in seconds, using the ″ symbol.) Also, if the object is oblong, the PA of the major axis will be shown.

The *class* of deep-sky object will also be shown.

For galaxies, two classification schemes are used—the older Hubble Scheme [in which S is spiral, SB is barred spiral, SO is spheroidal, E is elliptical, and Irr is irregular; subclasses designate the amount of arm winding (for spirals) or oblateness (ellipticals and spheroidals)] and the newer de Vaucouleurs scheme in which the following complex taxonomy is used:

Ellipticals (Hubble E)
     Compact types
         Hubble E6   cE
         Hubble E5   E0
         Hubble E5 intermediate   E0-1
         Hubble E4   E+

Lenticulars (Hubble S0)

Nonbarred   SA0

Barred   SB0

Mixed   SAB0

  Inner ring   S(r)0

  S-shaped   S(s)0

Mixed   S(rs)0

  Early (Hubble S03)   S0−

  Intermediate (Hubble S02)   S0°

  Late (Hubble S01)   S0+

Spirals (Hubble S)

Nonbarred   SA

Barred   SB

Mixed   SAB

  Inner ring   S(r)

  S-shaped   S(s)

  Mixed   S(rs)

Stages

  Hubble S0   SO/a

  Hubble S1   Sa

  Hubble S2   Sab

  Hubble S3   Sb

  Hubble S4   Sbc

  Hubble S5   Sc

  Hubble S6   Scd

  Hubble S7   Sd

  Hubble S8   Sdm

  Hubble S9   Sm

Irregulars (Hubble Irr)

Nonbarred   IA

Barred   IB

Mixed   IAB

  S-shaped = I(s)

  Non-Magellanic   I0

  Magellanic   Im

  Compact   cI

Peculiars   Pec

For all types, the following addends apply:

:   uncertain

?   doubtful

s   spindle

(R)   outer ring

(R')   pseudo-outer ring

The David Dunlop Observatory Spiral Luminosity type is then attached as a Roman numeral: from I = thick; well-developed arms down to V = anemic; poorly developed arms.

For *open clusters*, I use the Trumpler taxonomy as described in the Lick Observatory Bulletin, Number 14, page 154, 1930 issue. A Roman numeral (I to IV) indicates the concentration of the cluster from I (very concentrated) to IV (not well detached from the surrounding star field). A numeral from 1 to 3 indicates the range in brightness, with 1 being a small range and 3 a large range. Finally, a lower case letter indicates richness, with *p* being poor (fewer than 50 stars), *m* being moderately rich (50–100 stars), and *r* being rich (over 100 stars). In addition, a suffix of "n" means nebulosity is present.

Below, I give some examples of the Trumpler taxonomy using images from the Palomar Observatory Sky Survey. (I show negatives as more detail can be seen on a negative than on the print itself.)

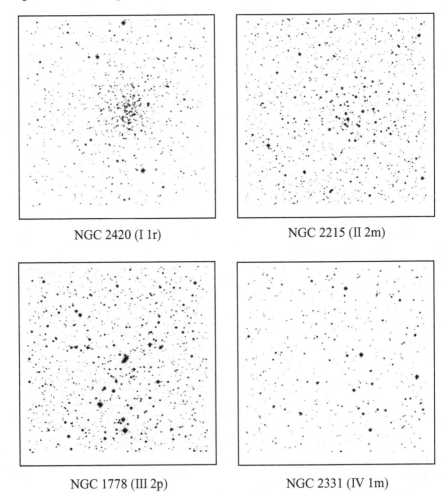

NGC 2420 (I 1r)   NGC 2215 (II 2m)

NGC 1778 (III 2p)   NGC 2331 (IV 1m)

For *globular clusters*, I use the taxonomy developed by Harlow Shapley and Helen Sawyer in the *Harvard Observatory Bulletin*, No. 824, 1927. Values range from 1 to 12 with the smaller the number indicating the more highly concentrated the stars are toward the center.

For *planetary nebulae*, I use the *Vorontsov-Velyaminov scheme* (1934), where the types range from I (stellar) to IIa (oval, evenly bright, concentrated), IIb (oval, evenly bright, not concentrated), IIIa (oval, unevenly bright), IIIb (oval, unevenly bright, brighter edges), IV (annular), V (irregular), and VI (anomalous).

In some cases, I will also show the old (and obsolete) *Herschel numbers* for double stars and deep-sky objects. For double stars, the Herschel catalog had these classes: I (difficult), II (close but measurable), III ($5''$ to $15''$), IV ($15''$ to $30''$), V ($30''$ to $60''$), VI ($1'$ to $2'$), and N (the 1821 catalog).

For deep-sky objects, Herschel's catalog used a similar system, but with different definitions for the codes: I (bright nebulae), II (faint nebulae), III (very faint nebulae), IV (planetary nebulae), V (very large nebulae), VI (very compressed rich clusters of stars), VII (compressed clusters of stars), and VIII (coarse clusters of stars).

Herschel numbers are no longer used by astronomers but are included for those who are pursuing their Herschel 400 and Herschel II awards from the Astronomical League.

The Deep-Sky Nomenclature

The deep-sky object nomenclature follows the code as given next:

3C   Third Cambridge Catalog of Radio Sources (nebulae, galaxies)

Barnard   E. E. Barnard (various objects)

Basel   (Open clusters)

Berkeley   (Open clusters)

Blanco   (Open clusters)

Bochum   (Open clusters)

Cederblad   (Nebulae)

Collinder   Per Collinder (Open clusters)

Czernik   (Open clusters)

Dolidze   (Open clusters)

Frolov   (Open clusters)

Haffner   (Open clusters)

Herschel   William Herschel (various objects)

King   (Open clusters)

M   Charles Messier (various objects)

Markarian   (Various objects)

MCG   Morphological Catalog of Galaxies

Melotte   (Open clusters)

NGC   New General Catalog (Dreyer), 1895, 1908 (various objects)

Palomar   (Globular clusters)

PK   Perez-Kohoutek (planetary nebulae)

Roslund   (Open clusters)

Ruprecht   (Open clusters)

Stephenson   (Open clusters)

Stock   (Open clusters)

Tombaugh   Clyde Tombaugh (globular and open clusters)

Trumpler   (Open clusters)

UGC   Uppsala General Catalog, 1973 (galaxies)

CHAPTER SIX

Framing the Picture

About the "Scale Models"

As an active member of the Astronomical Society of Kansas City, I often assisted our Society at public star parties and observing events, such as the 2003 Opposition of Mars and the appearance of various comets. Often, after the initial wave of people who would peer through our telescopes at the featured event were satisfied, we would then show them other celestial show stoppers. Often people would ask about how big a certain thing was and in an effort to accommodate their curiosity and impress them with the size of the universe, I developed a scale model system that usually left people stunned when I shared the models with them.

For example, when I would show someone M13, the Great Hercules Globular Cluster, they would do "Oooh!" and "Ahhh!". I would then share with them that if each of those stars (all of which are larger than the Sun to be visible at this distance) were grains of sand, the grains would be, on average, 3 miles apart and the entire cluster the size of Kansas, with about 1 million grains of sand being used. The usual reaction was a shaking of the head and a muffled "Wow!"

But to build good models, we must work within certain tight requirements. First, it is only possible to build scale models of objects (binaries, star clusters, etc.) when we know the distance (with some degree of accuracy) and the angular size as measured in a telescope. In the early 1990s, the European Space Agency (ESA) conducted a milestone mission named "Hipparcos" (and a secondary mission named "Tycho"), using orbiting observatories to collect positional data on millions of stars with unprecedented accuracy. A spin-off of this research was the discovery of thousands of new (and much more accurate) parallaxes to nearby stars.

The Hipparcos and Tycho data on parallaxes is very good out to about 500 light years from earth; beyond that range the accuracy begins to drop off rather quickly. So for double stars within that range, the models that I put forth are probably fairly accurate; beyond that distance, the model should be taken with a healthy allowance for variance.

You should also bear in mind that for the vast majority of the binary star models, the separations are projections only. That is, since we do not know the orbital parameters on most binary stars, we do not know how much the orbital plane is tilted with respect to our line of sight. Hence, the separation between two stars is the projection on the sphere of the sky of their line-of-sight positions, not necessarily their true positions. So unless we know the true orbit of a binary, the model will represent a *minimum* distance between the components. Once we determine the orbit to a given binary, the actual distance of separation will probably become greater. Also, since binary star systems are dynamic, the model is sized for the current observational epoch.

In modeling double stars, I use one of two methods to determine the size(s) of the stars for the model. If the complete spectral class is known (Morgan-Keenan type plus luminosity class, such as G5V), I make the assumption that the star is a typical representative of that class. We have a fairly good idea of how large different stars are based on their spectral types, so this is a reasonable approach to take. However, it is not 100% accurate for the same reason that we cannot always say that a 45-year old American male with a height of 6 feet will weigh exactly 168 lb. Some males of that age and size *will* weigh 168 lb, but many will not.

The other method to determine stellar sizes is to use the relationship between surface temperature and apparent magnitude. If we know the surface temperature (which is a direct function of the spectral class) and magnitude, along with distance, we can compute the star's luminosity. This is also a good approach, but allowances must be made for interstellar absorption of light (which is not always known), as well as the *effective* temperature of the star (roughly, it is "black body" temperature—the temperature it would be if it were a perfect radiator of light, which no star is). Once we know the effective temperature and magnitude, we can compute how many lumens per square foot the star's surface emits and deduce its size if we know its distance. In most cases we do not know the effective temperature precisely, so we assume it is close to the spectral temperature. Like the spectral class model approach, this approach is not perfect either, but it gives us a fairly good idea of how star sizes vary.

There are very few deep-sky objects within 500 light years of earth, so the distances to deep-sky objects (open clusters, globular clusters, planetary nebulae, and so on) are often just good estimates. In a few cases, Cepheid variable stars are present in these objects that allow us get a rather accurate fix on their distances, but for the most part, distances to deep-sky objects are, at best, educated estimates.

In my models, I have chosen to let the Sun be represented by a 3-inch (7.62 cm) sphere—roughly, the size of an American baseball. On this scale, the earth would be 0.0274 inches (0.6 mm) in diameter—about the size of a ball point pen's roller ball. This ball would be almost 27 feet (8.19 m) away from the baseball. Pluto would lie about 1062 feet (324 m) away. *A light year, at this scale, would be about 322 miles (518 km).*

Astronomers generally agree that for binary stars in the galactic disc, if the stars are more than a light year apart, they are probably not truly binary as the local tugs on the pair will pull them from each other's grasp in less than one galactic orbit; so when my models show a component to be more than 322 miles (518 km) away from its primary, more than likely, we are not looking at a true binary star system. But the final verdict may take centuries to reach as enough of the companion's motion can be measured to decide if it is on an orbital path or a passing (hyperbolic) path.

Finally, the models as stated in this study all use English systems of measurement (inches and miles). For those who live in Metric system-based cultures, inches may be converted to centimeters by multiplying the inches by 2.54, feet to centimeters by multiplying the feet by 30.5, yards to meters by multiplying the yards by 0.91, and miles to kilometers by multiplying the miles by 1.61. Thus, the Sun in a Metric scale would be about 7.62 cm in diameter, while a light year (322 miles) would be about 518 kilometers.

Any references to luminosity of an object (stated in Solar equivalents) is somewhat approximate as well. For instance, it is not always easy to determine how much interstellar absorption occurs as light from a distant source approaches earth, and even a small change in the transmission of light can make a substantial difference in the calculated luminosity. For galaxies, the luminosity is based on the object's magnitude (never a precise measure for extended objects like galaxies) and the aspect that it presents to us. For instance, if a spiral galaxy is highly tilted away from our line of sight (so that it appears as a spindle or discus), the luminosity computed for that galaxy is based on the light that we see—the actual amount of light being emitted by the galaxy would be much higher as a face-on observer would catch the galaxy in nearly full-illumination mode (with something over half the actual output of the galaxy reaching the observer's eye, the other fraction being emitted on the opposite side of the galaxy). So take the galaxy luminosities with a fair amount of skepticism, allowing for all these variables that make computing a true luminosity a most difficult task.

Frequent Sketches

The Complete CD Guide to the Universe contains about 800 sketches I have made at the eyepiece. These sketches were made with a soft pencil and smudging stick (where appropriate) and scanned into a graphic file format and inserted in the zone catalogs where appropriate. These sketches appear just as I made them at the eyepiece and will usually include the date of the sketch and the sky conditions. You will see things like "s4, t3." This means that on that night, the seeing (steadiness of the air) was 4 out of 5 (very good), while the transparency (clearness of the air) was 3 out of 5 (about average).

CHAPTER SEVEN

Maps and Zones

A Note About Map Scales

There are many different scales of maps used in this book. The largest scale is for an entire zone and is approximately 10 degrees square. The faintest stars plotted on this map are magnitude 9.0.

The detail maps come in two versions—a normal view (as the sky would look to the eye without inversion or reversal) and a mirror image view (upright but mirrored). Between these two chart options, you can have an eyepiece view for any telescope (by either using the charts as printed or by rotating the page). These detailed charts show stars to 12.5 magnitude and deep-sky objects to 15.0 magnitude. Scales vary depending on how much I had to zoom in to show enough detail, but generally most zones are divided into four subzones, meaning the typical subzone map is about 5 degrees square. Some detail maps go down to as small as 2 degrees square. Since each zone map has an equatorial grid overlaid on it, you can determine the map scale from the grid without too much difficulty.

It is all About Zones

The entire book is set up using Zones. A zone is 1-hour-angle wide and 10 degrees tall (except near the poles, where the zone may span 3 hour angles).

Each *zone file* will open with a full zone map, showing the orientation and numbering of the detailed finder maps. The finder maps will then be displayed after the zone map in both normal and mirror modes.

Each zone file also has a *zone catalog* that describes all the objects I have observed in that zone.

The object descriptions will be set up in two areas. The first area will be the description of the double stars on the finder map. The second section will be a description of the deep-sky objects on the map.

In each section, the objects will be presented in order of increasing difficulty and decreasing "rating." I use three *difficulty levels*—easy, moderate, and difficult. These are my way of telling you what you can expect under ordinary seeing conditions.[1] Bear in mind that most of my observations have been made from a suburban location, not the darkest or cleanest skies in America. So when I say a particular galaxy is moderate in difficulty, that is my rating from my light-polluted location (unless otherwise noted). If you are lucky enough to live in an area with darker skies, my "moderate" may be your "easy" and so on. Newcomers to this wonderful hobby may want to start out trying to observe all of the "easy" objects in a zone, then working up to the "moderate" ones, and as skills improve, try for the "difficult" ones.

The *rating* is a scale I developed early in my observing career when I began logging my observations in a database on my computer. I wanted some way to rapidly query the database for various types of objects that I could show a guest on short notice. Let us face it—you and I, as experienced observers, would have fun observing almost anything with our telescopes. But to the layman who has not spent any time at a telescope, dim and difficult double stars or BARFS (big and really faint stuff) would probably be boring beyond belief. So I began rating each object with a scale of 1 to 5, where 1 is a stunning, wonderful view and 5 is a view that is hardly worth the effort to get.

So combining the difficult and rating scales, you will have an idea of what you might expect of any object before you see it. A double star, for instance, rated a 5E, would be very easy but awfully boring. (For example, it may consist of two 7th magnitude stars 180" apart a split an elephant could waltz through!) Likewise, a galaxy rated 5D would be boring but awfully difficult. (That is not to say that seeing ancient light is boring, but after viewing thousands of galaxies, I can say that many of them are just about as exciting to watch as drying paint.)

---

[1] About 95% of the objects in this study have been observed from urban settings; about half of them with a C-8 from the outskirts of Columbia, MO, form 1987 to 1990, and the other half from suburban north Kansas City, MO, from 1990 to the present, using the C-8 (up to 2002) and the C-11 since 2002. In the few cases where I could only observe an object from extra dark skies or with different instruments, this will be noted in the catalog description. My point is that there is a great deal of observational astronomy that can be done even from a moderately light-polluted area if you are patient and develop your skills.

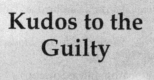

CHAPTER EIGHT

Kudos to the Guilty

Sources and Acknowledgments

A study of this size and scope is the result of the efforts of many people and organizations and as Isaac Newton once quipped to a colleague, if I have seen farther than other people, it is because I have stood on the shoulders of giants. It is therefore fitting that I give full recognition to all those who have helped in the development of my hobby and the preparation of this book.

First and foremost, all the star charts in this book were generated with *TheSky* astronomy software, version 5.0 by Software Bisque, copyright Software Bisque, Inc. All rights reserved. www.bisque.com.

I currently use version 6.0, but used 5.0 to create this book since 6.0 came out late in the writing. I will be upgrading to Bisque's newest version as soon as it is available.

Star notes/essays (referred to in the text as *James Kaler's Star Notes*) are from http://www.astro.uiuc.edu/~kaler/sow/sow.html, copyright James B. Kaler, Department of Astronomy, University of Illinois. Used with permission. Sources Dr. Kaler credits are

Allen, R. H. *Star Names: Their Lore and Meaning.* Dover, New York, reprint, 1963.

The Astronomical Journal.

The Astronomical Society of the Pacific Conference Series.

Astronomy and Astrophysics: A European Journal.

Astronomy and Astrophysics Abstracts. Springer, Heidelberg.

Astronomy and Astrophysics Supplement Series.

The Astrophysical Journal.

The Astrophysical Journal Supplement Series.

Berry, A. *A Short History of Astronomy.* Dover, New York, reprint, 1961.

Hoffleit, D. and Jaschek, C. *The Bright Star Catalogue,* 4th ed. Yale University Observatory, New Haven, CT, 1982.

The Hipparcos and Tycho Catalogues. European Space Agency, Peris, France 1998.

Kaler, J. B. *Astronomy!* HarperCollins, New York, 1994.

Kaler, J. B. *Stars and their Spectra.* Cambridge University Press, Cambridge, Massachusetts, 1989.

Kunitzsch, P. and Smart, T. *Short Guide to Modern Star Names/Derivations.* Harrassowitz, Berlin, 1986.

Monthly Notices of the Royal Astronomical Society.

The Observatory. Royal Astronomical Society, Grove, Wantage, UK

Bishop, R. (ed.) *Observer's Handbook.* Royal Astron. Soc. Canada, Toronto, U. Toronto Press.

Publications of the Astronomical Society of the Pacific.

SIMBAD on the Web, CDS, University of Strasbourg.

Tirion, T. *Sky Atlas 2000.* Sky Publishing Corp., Cambridge, 1981.

Other public domain sources include

Astronomy Magazine, numerous issues from 1975 to 1996.

Burnham, R. *Burnham's Celestial Handbook* (Vol. 1–3). Copyright 1968, 1978 by Dover, New York, NY.

Catalogue of Open Cluster Data. Downloaded from CompuServe's Astronomy Forum in the early 1990s.

The Centre de Données stronomiques de Strasbourg, which runs an awesome web site (http://simbad.u-strasbg.fr/sim-fid.pl).

Reverend Webb, T. W. *Celestial Objects for Common Telescopes.* Copyright 1962 by Dover, New York, NY. Webb's original book was published in 1917.

Mitton, J. *A Concise Dictionary of Astronomy.* Copyright 1991 by Oxford University Press, Oxford, England.

Illingworth, V. (ed.) *The Facts on File Dictionary of Astronomy,* 3rd ed. Copyright 1979, 1985, 1994 by Market House Books, Ltd. Published by Facts on File, Inc. New York, NY.

Cragin, M., Lucyk, J., and Rappaport, B. *The Deep Sky Field Guide to Uranometria 2000.0.* Copyright 1993 by Willmann-Bell, Inc. Richmond, VA.

The New General Program, a computer database and record-keeping system for deep sky observations by G. Dean Williams. Downloaded in 1994 from CompuServe's Astronomy Forum, along with observational notes uploaded by Dean Williams, Barbara Wilson, Eric Greene, Steve Coe, and Geoff Chester, as well as files from The Morphological Catalog of Galaxies, the Lynds Catalog of Bright Nebulae, and The Lynga Catalog of Open Clusters.

RealSky, the Palomar Observatory Sky Survey on CD-ROM, with the Equatorial and Southern Sky Surveys of the Siding Spring Observatory, Australia. Joint project of the Astronomical Society of the Pacific and the Space Telescope Science Institute, cosponsored by The Association of Universities for Research in Astronomy, Inc. (AURA).

Humason, M., Mayall, N., and Sandage, A. Redshifts and magnitudes of extragalactic Nebulae. *The Astrophysical Journal* (Vol. 61, Reprint No. 181) pp. 97–162, 1956.

The Saguaro Astronomy Club (SAC) Database, by members of the Saguaro Astronomy Club of Phoenix, AZ. A special thanks to Steve Coe and A. J. Crayon for all their help and assistance!

Hirshfeld, A., Sinnott, R. W., and Ochsenbein, F. *Sky Catalog 2000.0* (Vol. 1 and 2). Copyright 1991 by Sky Publishing Corporation, Cambridge, MA, and Cambridge University Press, Cambridge, MA.

Sky and Telescope Magazine, numerous issues from 1994 to the present.

Jones, K. G. (ed.) *The Webb Society's Deep Sky Observer's Handbooks* (Vol. 1–8). Compiled by members of The Webb Society. Copyright 1975 by The Webb Society. Published by Enslow Publishes, Inc. Hants, UK.

The Washington Double Star Catalog, compiled by Brian Mason, and downloaded from the US Naval Observatory's web page.

The Yale Bright Star Catalog, downloaded from www.pomona.claremont.edu.

All Hubble Space Telescope pictures used for the CD Menus are used with the permission of National Aeronautics and Space Administration (NASA) and the Space Telescope Science Institute (STScI).

Main Menu   Hubble Deep Field. Image credit NASA, ESA, F. Summers (STScI).

Maps and Catalogs, 00–05 h   NGC 1850. Image credit NASA, ESA, Martino Romaniello (ESO).

Maps and Catalogs, 06–11 h   Orion Nebula. Image credit NASA, ESA, The Hubble Heritage Team (STScI/AURA).

Maps and Catalogs, 12–17 h   M104. Image credit NASA and the Hubble Heritage Team (STScI/AURA).

Maps and Catalogs, 18–23 h   M17. Image credit NASA, ESA and J. Hester (ASU).

Support Files Menu   Supernova Remnant N 63A. Image credit NASA, ESA, HEIC, The Hubble Heritage Team (STScI/AURA).

Index Menu   Supergiant V838 Mon. Image credit NASA, ESA, H. E. Bond (STScI).

Sky Albums Menu   Quintuplet Cluster. Image credit NASA, ESA, STScI.

CHAPTER NINE

How to Use the CD-ROM

The observing guide is entirely contained on one CD-ROM. All the files on the CD are prepared with Word and graphics programs, then converted into portable document files (PDFs) using Adobe Acrobat.

The CD was created with the intention of the observer selecting a zone (or zones) to observe and then printing what information he or she wanted to assist in observing that zone. *It is not well suited for use at the telescope on a laptop computer* simply because Adobe Acrobat Reader, the program that lets you view the maps and object descriptions, does not work well in the night vision modes of most astronomy programs. The window and tool bars will be reddened, but the page will still be a brilliant white. You can manually create a display profile in Windows that uses red background for the white pages and a white or gray text for the printing. I leave it to the reader to discover how to do this on their own or to seek assistance from a computer-savvy helper.

Starting the CD

To use the CD, you will have to have Adobe Acrobat Reader, version 5.00 or later, installed on your computer. If you do not have Acrobat Reader installed, you may also use the Internet and download the latest Reader version free from Adobe. Navigate to www.adobe.com and follow the links to Acrobat Reader.

Once you have installed Adobe Acrobat Reader, you are ready to use the CD. To start the CD, launch Acrobat Reader. Then use the file commands (File, Open), or click on the *File Open* tool icon (![Open]( )) and set the open dialog box to find the CD. Your dialog box should look something like as given next.

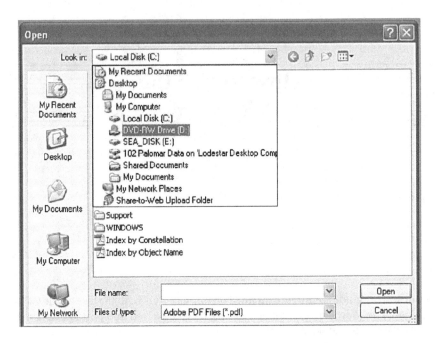

Set the location to the CD and the dialog box will change to look like as follows.

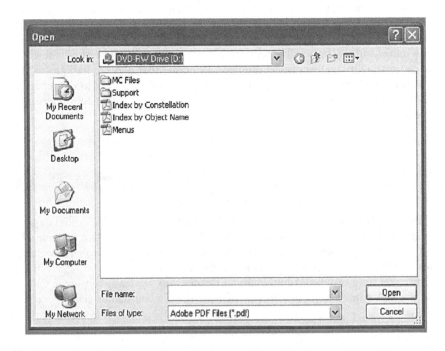

Find the file named Menus and open it. You are now up and running.

Alternative Launch Method

If your computer is set to display a CD contents automatically upon loading the CD, you will find a window similar to the one shown just above on your screen. Merely double-click the Menus file to start *The Complete CD Guide to the Universe.*
The *Main Menu* looks like as follows:

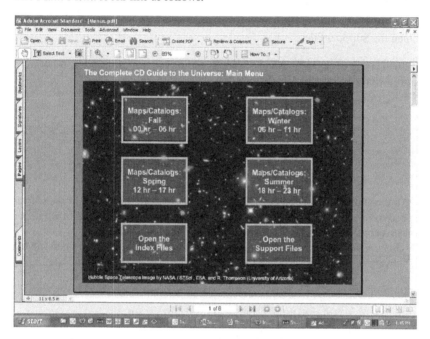

You may use Acrobat Reader's zoom command to change the sizing of the menu on the screen, or you may also click the page display tools to change how the menu program displays the opening menu, but the default is probably the best all-around display for starters.

The main menu uses six icons to take you to various files in *The Complete CD Guide to the Universe.* To go to one of the menu options, merely hover your mouse over it and then click it.

Note that while the mouse is *not* over a menu option, the icon for the mouse is a small hand, like as follows:

And when the mouse is over a menu option, it changes to a pointing finger (see next).

The Main Menu Icons

There are six options on the *Main Menu*. They are

1. Maps/Catalogs (Fall): 00–05 h
2. Maps/Catalogs (Winter): 06–11 h
3. Maps/Catalogs (Spring): 12–17 h
4. Maps/Catalogs (Summer): 18–23 h
5. Open the Index Files
6. Open the Support Files

We will come back to the Maps/Catalogs options in a moment. For now, we will focus on what the 5th and 6th options do.

Finding a Particular Object: The Index Files

There are roughly 13,200 objects detailed in *The Complete CD Guide to the Universe*. Suppose you wanted to see if a particular object was covered, and if so, in what Zone? That is where the *Index Files* icon on the *Main Menu* comes into play.

By clicking on the *Open the Index Files* icon on the *Main Menu*, you open the *Index Files* submenu.

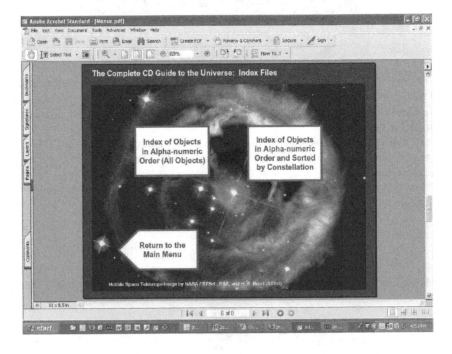

You have two options available to you—you may search by object name or for all the objects in a constellation.

Let us suppose you wish to observe NGC 1003. Is it covered in *The Complete CD Guide to the Universe,* and if so, in what Zone?

Click on the *Index of Objects in Alpha-Numeric Order* icon to load that Acrobat file. The first page of that file looks like as follows.

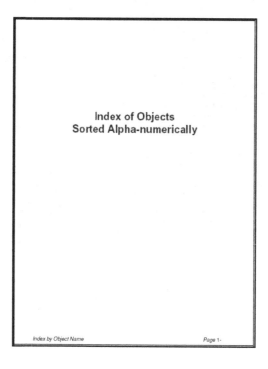

Once the file is loaded, search for NGC 1003. For me, the easiest way is to press *Control* and *F* at the same time (or click on the *Binoculars* tool) and type NGC 1003 in the *Search for* window. Then click *Search* and wait while Acrobat looks for the text string. (You will find NGC 1003 in Zone 84, Map 3.)

Now suppose you wanted to see if NGC 2412 was covered. Do a search for NGC 2412. You will find that it is not covered in this work. (It is classed as "non-existent" in the revised NGC.)

Suppose now you wanted to see what is covered in Cancer, the Crab. You would click the *Objects Sorted by Constellation in Alpha-Numeric Order* icon to open the constellation index file.

What Is in the Support Files?

Clicking on the *Open the Support Files* icon on the *Main Menu* opens the *Support Files* submenu. It looks like as follows:

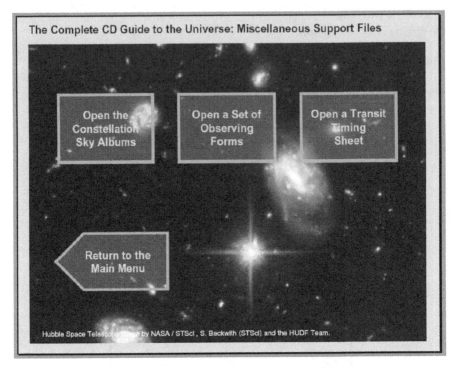

This menu offers a rich treasury of files to help increase your observing efficiency and pleasure.

We will begin by examining what the *Open a Set of Observing Forms* icon does. It opens a PDF file which consists of a four-page document I downloaded from The Saguaro Astronomy Club's web page (www.saguaroastro.org). I have found no better field observing forms than these and strongly recommend them. You may print these off in whatever quantity you need and whichever version you like most for use at the telescope. (In Acrobat Reader, you may print an entire document or any page or combination of pages; refer to the *Acrobat Reader help* file if you are not certain how to do this.)

The *Transit Timing Form* is a PDF document that helps you determine the field of view of any eyepiece you have.

The Sky Albums

The icon labeled *Open The Constellation Sky Albums* opens a new submenu that allows you access to any constellation's sky album. The submenu looks like as shown next:

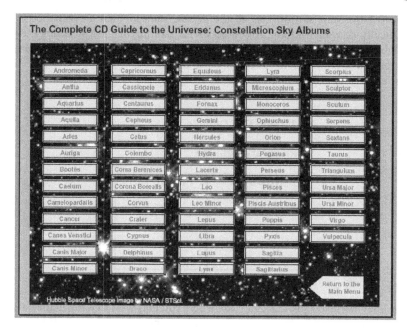

The Complete CD Guide to the Universe: Constellation Sky Albums

Andromeda	Capricornus	Equuleus	Lyra	Scorpius
Antlia	Cassiopeia	Eridanus	Microscopium	Sculptor
Aquarius	Centaurus	Fornax	Monoceros	Scutum
Aquila	Cepheus	Gemini	Ophiuchus	Serpens
Aries	Cetus	Hercules	Orion	Sextans
Auriga	Colombo	Hydra	Pegasus	Taurus
Boötes	Coma Berenices	Lacerta	Perseus	Triangulum
Caelum	Corona Borealis	Leo	Pisces	Ursa Major
Camelopardalis	Corvus	Leo Minor	Piscis Austrinus	Ursa Minor
Cancer	Crater	Lepus	Puppis	Virgo
Canes Venatici	Cygnus	Libra	Pyxis	Vulpecula
Canis Major	Delphinus	Lupus	Sagitta	
Canis Minor	Draco	Lynx	Sagittarius	

Return to the Main Menu

Hubble Space Telescope Image by NASA / STScI

So what is a *Sky Album*?

When I was a lad, I had a stamp collection. It was nothing to turn the head of an investment collector, but I was proud of it nonetheless. I had purchased one of Scott's largest albums and was dutifully acquiring stamps that would cover the half-tone images of the stamps that were issued by that nation as I affixed them to the album with stamp hinges. I could tell at a glance which stamps I still needed to acquire in order to complete an issue.

That is the concept of a Sky Album. For example, a part of the album for Andromeda appears next.

1. Double Stars

Difficulty Class: Easy

STF 3042	STF 24	OS 547	Gamma And	STF 17
Zone: 105, Map: 4	Zone: 106, Map: 2	Zone: 58, Map: 2	Zone: 84, Map: 1	Zone: 106, Map: 2
Rating: 1	Rating: 1	Rating: 1	Rating: 1	Rating: 2
2352+3753	0018+2608	0006+4548	0204+4200	0017+2918
5.7 +" @ 87-	5.0 +" @ 250 +	6.1 +" @ 182 +	10 =" @ 63 +	26.9 +" @ 30 +
Obs: __/__/__	Obs: __/__/__	Obs: __/__/__	Obs: __/__/__	Obs: __/__/__
STF 28	**STF 3**	**h2010**	**STF 102**	**STF 2985**
Zone: 106, Map: 2	Zone: 58, Map: 2	Zone: 59, Map: 2	Zone: 59, Map: 2	Zone: 81, Map: 3
Rating: 2	Rating: 2	Rating: 2	Rating: 2	Rating: 2
0024+2930	0010+4623	0102+4742	0118+4900	2310+4757
33.0 +" @ 224+	5.0 +" @ 82-	9.9 =" @ 270-	0.5 =" @ 277-	15.6 +" @ 256 +
Obs: __/__/__	Obs: __/__/__	Obs: __/__/__	Obs: __/__/__	Obs: __/__/__
A 799	**OS 514**	**STF 79**	**STF 44**	**STF 154**
Zone: 81, Map: 8	Zone: 12, Map: 1	Zone: 82, Map: 3	Zone: 82, Map: 3	Zone: 83, Map: 3
Rating: 2	Rating: 2	Rating: 2	Rating: 2	Rating: 2
2358+4804	0005+4205	0100+4442	0038+4059	0145+4342
2.0 +" @ 11-	5.3 +" @ 169-	8.3 +" @ 192-	12.2 +" @ 273+	5.4 +" @ 126+
Obs: __/__/__	Obs: __/__/__	Obs: __/__/__	Obs: __/__/__	Obs: __/__/__
STE 470	**STE 245**	**50 And**	**STE 2002**	**A C 301**

The Double Stars section lists all the double stars for that constellation sorted by difficulty class then rating. Any object with a rating of 1—a stunning view—is highlighted in yellow (and will print in yellow if you have a color printer; otherwise, it will print as a gray tone, like the samples in this book).

The same general format applies for deep-sky objects.

2. Deep Sky Objects

Planetary Nebulae

NGC 7662	PK 107-13.1
Zone: 11054, Map: 1	Zone: 81, Map: 6
Rating: 2E	Rating: 4
Mv = 8.9	Mv = 13.9
Obs: —/—/—	Obs: —/—/—

The lines marked "Obs: __/__/__" are where you may wish to write in the date you observed that object as you attempt to collect all the "stamps" that will fill in your "album."

The Heart of the System: The Maps and Catalogs

There are four similar-looking icons on the *Main Menu*. Since each works the same way, we will spend time with only one of them.

When you click this icon (or any of its "siblings"), a special submenu opens. The submenu for the first group of Maps and Catalogs is as follows:

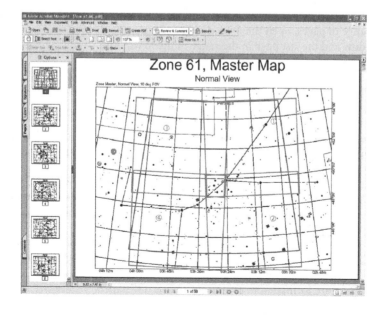

At the top of each *Maps and Catalogs* submenu is a small button. It takes you back to the *Main Menu.*

The other buttons open the various Zone files.

As an example, let us suppose that this evening you wanted to do some observing in the portion of the sky bounded by 03 h to 04 h and +45° to +55°. A quick check of the menu shows that you need to open Zone 61.

Clicking the Zone 61 button starts Adobe Acrobat Reader and displays the opening view as shown next:

Note how on the left you will see a series of miniaturized pages ("thumbnails") with the current page outlined with a *bold* line.

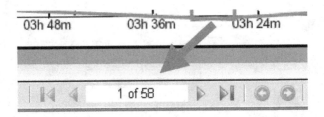

You may quickly go to any previewed page be merely clicking on its thumbnail. You may also go directly to a page by clicking in the pages window at the bottom of the screen and typing the page number, then pressing Enter. The small triangles that flank the page number window let you also browse up or down one page at a time, or jump to the last page or first page (the triangles with the bars).

You can also change the "zoom" factor, which magnifies or reduces the page image in the page window.

You may print all or any part of a file. To print the entire file, merely click on the *Print* icon on the Reader toolbar. To print a specific page, go to that page, then click *File*, then *Print*. In the dialog box that opens, click *Current Page*, then *OK*. To print a specific range of pages, open the *Print* dialog and enter the starting page in the *Pages from* window and the last page in the range in the *to* window.

Saving Paper

Zone 61 contains 58 pages (and many Zones are even larger than that). You may not want to print that much material for use at the telescope. There are several ways to lighten your load.

First, just print the maps you *need*. For instance, I doubt if you will need the *Zone Master Map* unless you want it for star hopping. But you will need some of the detailed maps. Choose the map orientation that is best for your telescope and print *only* those maps.

As for the Catalog data, if you do not want all that information with you at the telescope, you will find *Mini-Catalogs* of all the objects in that Zone at the end of the catalog in a condensed tabular format.

All PDF files are formatted to print on 8-/12″ × 11″ paper.

Saving Time

If you are just going to be looking at one zone during your session, you may simply close Acrobat Reader when done (by clicking the red X in the far upper right of the Reader window). This action closes the Zone file and Acrobat at the same time.

But if you want to view other files during your session, close the document while leaving Acrobat open. You can do this by either pressing Control and W at the same time or by clicking the blue X below the red Acrobat Close X on the screen.

See the next picture to see what these *X* icons look like.

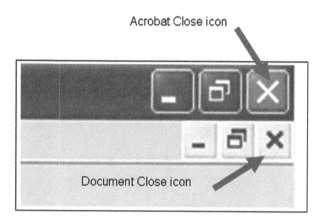

Acrobat Close icon

Document Close icon

The Appendix contains the entire file for one of the Zones to give you an idea of what a complete printout looks like.

Enjoy your journey! Take your time! There is enough in this book to keep you busy the rest of your life, and I am giving out no prizes for getting done first.

So go out under the stars and observe all you want. Enjoy your universe. The tour from your observing site is free.

Well, okay, you have to buy a telescope and some accessories, I suppose, so it is not really free, but . . . you get the idea! Clear and steady skies to you all!

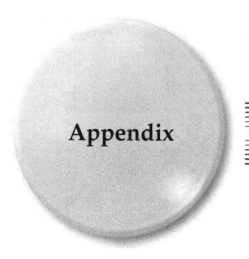

Appendix

Table A.1. Object Count

Object Type	Count
Double stars	10,738
Dark Nebulae	4
Bright Nebulae	68
Planetary Nebulae	132
Supernova Remnant	6
Open clusters	486
Globular clusters	104
Galaxies	1,458
Quasi-stellar Objects	1
Reference stars, red stars, dwarfs, etc.	241
Total	13,238

Table A.2. Page Counts for the Map and Catalog PDF Files

Zone Number	Pages of Maps	Pages of Catalogs	Total Pages
1	15	22	37
2	11	27	38
3	11	26	37
4	7	11	18
5	9	11	20
6	9	17	26
			(cont.)

Table A.2. (*Continued*)

Zone Number	Pages of Maps	Pages of Catalogs	Total Pages
7	9	18	27
8	9	13	22
9	11	21	32
10	11	21	32
11	7	20	27
12	9	16	25
13	5	25	30
14	5	18	23
15	5	16	21
16	5	12	17
17	5	11	16
18	5	14	19
19	5	13	18
20	5	9	14
21	5	13	18
22	7	18	25
23	5	9	14
24	5	13	18
25	5	7	12
26	9	12	21
27	5	13	18
28	9	12	21
29	7	15	24
30	3	5	8
31	7	19	26
32	9	32	41
33	13	24	37
34	25	92	117
35	43	85	128
36	21	63	84
37	9	43	52
38	9	37	46
39	9	20	29
40	9	23	31
41	9	22	31
42	9	16	25
43	7	13	20
44	9	14	23
45	9	32	41
46	9	19	28
47	9	16	25
48	9	22	31
49	9	21	30
50	9	19	28
51	9	21	30
52	9	26	35
53	9	31	40
54	9	33	42
55	11	46	57
56	19	72	91

Table A.2. (*Continued*)

Zone Number	Pages of Maps	Pages of Catalogs	Total Pages
57	17	64	81
58	9	58	67
59	9	45	54
60	9	36	45
61	9	46	55
62	9	41	50
63	9	40	49
64	9	29	38
65	9	25	34
66	9	31	40
67	9	16	25
68	9	17	26
69	11	26	37
70	11	25	36
71	9	22	31
72	9	34	43
73	9	19	28
74	9	24	33
75	9	28	37
76	9	22	31
77	19	45	64
78	19	54	73
79	19	60	79
80	19	66	85
81	19	51	70
82	9	46	55
83	9	45	54
84	11	55	66
85	19	62	81
86	9	32	41
87	9	54	63
88	9	44	53
89	9	27	36
90	9	27	36
91	9	19	28
92	9	22	31
93	9	29	38
94	9	40	49
95	9	34	43
96	9	25	34
97	9	28	37
98	9	32	41
99	9	25	34
100	15	47	62
101	21	70	91
102	19	95	114
103	9	60	69
104	9	55	64
105	9	35	44

(*cont.*)

Table A.2. (*Continued*)

Zone Number	Pages of Maps	Pages of Catalogs	Total Pages
106	9	42	51
107	11	40	51
108	9	37	46
109	9	38	47
110	9	20	29
111	15	62	77
112	9	37	46
113	9	38	47
114	9	34	43
115	9	24	33
116	9	25	34
117	9	20	29
118	9	36	45
119	9	25	34
120	9	30	39
121	9	25	34
122	9	24	33
123	9	35	44
124	15	62	77
125	19	73	92
126	9	92	101
127	9	35	44
128	9	38	47
129	9	30	39
130	9	27	36
131	9	27	36
132	9	23	32
133	11	29	40
134	9	39	48
135	9	48	57
136	15	61	76
137	9	47	56
138	11	26	37
139	9	20	29
140	11	33	44
141	9	31	40
142	9	26	35
143	9	23	32
144	9	24	33
145	9	19	28
146	9	22	31
147	11	34	45
148	9	45	54
149	19	66	85
150	9	54	63
151	9	29	38
152	9	26	35
153	11	33	44
154	9	20	29
155	11	35	46

Table A.2. (*Continued*)

Zone Number	Pages of Maps	Pages of Catalogs	Total Pages
156	9	24	33
157	9	16	25
158	9	30	39
159	9	40	49
160	21	64	85
161	9	37	46
162	9	35	44
163	11	23	34
164	15	35	50
165	9	37	46
166	19	74	93
167	11	27	38
168	11	25	36
169	9	23	32
170	9	28	37
171	9	28	37
172	9	52	61
173	9	56	65
174	9	50	59
175	9	38	47
176	9	25	34
177	13	35	48
178	9	28	37
179	19	38	57
180	9	31	40
181	9	25	34
182	9	29	38
183	21	71	92
184	21	65	86
185	19	62	81
186	9	37	46
187	9	25	34
188	9	23	32
189	9	24	33
190	11	42	53
191	9	25	34
192	9	29	38
193	9	23	32
194	9	30	39
195	9	36	45
196	9	44	53
197	9	42	51
198	9	32	41
199	9	31	40
200	9	37	46
201	9	22	31
202	9	21	30
203	9	27	36
204	9	28	37

(cont.)

Table A.2. (Continued)

Zone Number	Pages of Maps	Pages of Catalogs	Total Pages
205	9	24	33
206	9	28	37
207	9	53	62
208	9	49	58
209	19	58	77
210	9	31	40
211	9	22	31
212	9	20	29
213	9	17	26
214	9	28	37
215	9	21	30
216	9	22	31
217	9	19	28
218	9	21	30
219	9	42	51
220	9	37	46
221	9	28	37
222	9	21	30
223	9	20	29
224	9	20	29
225	9	20	29
226	9	21	30
227	9	18	27
228	9	14	23
229	9	23	32
230	9	15	24
231	9	27	36
232	9	39	48
233	21	62	83
234	9	25	34
235	9	22	31
236	9	24	33
237	9	23	32
238	9	20	29
239	9	22	31
240	9	20	29
241	9	23	32
242	9	21	30
243	9	39	48
244	19	73	92
245	9	26	35
246	9	30	39
247	9	20	29
248	9	23	32
249	9	17	26
250	11	16	27
251	9	9	18
252	11	12	23
253	11	14	25
254	9	14	23

Table A.2. (*Continued*)

Zone Number	Pages of Maps	Pages of Catalogs	Total Pages
255	7	12	19
256	11	25	36
257	13	42	55
258	13	31	44
259	11	17	28
260	9	14	23
261	11	18	29
262	11	14	25
263	13	22	35
264	11	22	33
265	13	25	38
266	13	37	50
267	13	50	63
268	13	41	54
269	11	17	28
270	11	24	35
271	9	13	22
272	7	9	16
273	9	16	25
Grand Totals	2,793	8,533	11,326

Table A.3. Page Counts for the Index and Support Files

Document	Pages
Index (alpha by object)	319
Index (by constellation)	311
Observing forms	4
Transit timing form	1

Table A.4. Page Counts for the Sky Albums

Constellation	Pages
Andromeda	17
Antlia	2
Aquarius	9
Aquila	12
Aries	5
Auriga	15
Boötes	11
Caelum	1
Camelopardalis	13
Cancer	7
	(cont.)

Table A.4. (*Continued*)

Constellation	Pages
Canes Venatici	7
Canis Major	8
Canis Minor	4
Capricornus	5
Cassiopeia	23
Centaurus	2
Cepheus	15
Cetus	11
Colombo	1
Coma Berenices	6
Corona Borealis	2
Corvus	3
Crater	3
Cygnus	25
Delphinus	4
Draco	14
Equuleus	2
Eridanus	8
Fornax	3
Gemini	11
Hercules	17
Hydra	13
Lacerta	8
Leo	11
Leo Minor	3
Lepus	5
Libra	5
Lupus	2
Lynx	8
Lyra	9
Microscopium	1
Monoceros	12
Ophiuchus	11
Orion	12
Pegasus	14
Perseus	15
Pisces	9
Piscis Austrinus	2
Puppis	8
Pyxis	3
Sagitta	4
Sagittarius	11
Scorpius	5
Sculptor	3
Scutum	3

Table A.4. (Continued)

Constellation	Pages
Serpens	7
Sextans	3
Taurus	12
Triangulum	3
Ursa Major	16
Ursa Minor	4
Virgo	17
Vulpecula	7
Total Pages	512

Table A.5. Distribution of Rating and Difficulty

Rating	Easy	Moderate	Difficult
1	135	14	9
2	567	141	98
3	1,744	499	264
4	2,822	1,009	630
5	1,963	1,485	1,362

Table A.6. The Messier Objects

Messier No.	Season	Zone	Map	Other Names
1	Fall	135	3	Crab Nebula; 3C144
2	Summer	199	4	
3	Spring	119	4	
4	Spring	266	1	
5	Spring	193	1	
6	Spring	267	5	Melotte 178; Collinder 341; Raab122; Butterfly Cluster
7	Spring	267	5	Melotte 183; Collinder 354; Raab125
8	Summer	244	3	Lagoon Nebula; Hourglass Nebula
9	Spring	243	1	
10	Spring	194	4	
11	Summer	220	3	Melotte 213; Collinder 391; TheWild Duck
12	Spring	194	4	
13	Spring	98	4	The Great Hercules Cluster
14	Spring	195	4	
15	Summer	175	3	
16	Summer	220	2	Melotte 198; Collinder 375; Eagle; Star Queen
17	Summer	244	4	Collinder 377; Omega Nebula; Swan Nebula
18	Summer	244	1	Collinder 376
19	Spring	267	1	
20	Summer	244	3	Trifid Nebula; H IV 41; Bochum 14

(cont.)

Table A.6. (Continued)

Messier No.	Season	Zone	Map	Other Names
21	Summer	244	3	Melotte 188; Collinder 363; Raab128
22	Summer	244	6	
23	Spring	243	3	Melotte 184; Collinder 356; Raab126
24	Summer	244	2	Melotte 197
25	Summer	244	5	Melotte 204; Collinder 469; Raab131
26	Summer	220	3	Melotte 212; Raab 136
27	Summer	150	1	Dumbbell Nebula; PK 60-3.1
28	Summer	244	6	
29	Summer	102	5	Collinder 422
30	Summer	247	4	
31	Fall	82	3	Great Andromeda Galaxy
32	Fall	82	3	
33	Fall	107	3	Pinwheel Galaxy
34	Fall	84	3	Melotte 17; Collinder 31; Raab 13
35	Winter	136	1	Melotte 41; Raab 31
36	Fall	111	3	Melotte 37; Collinder 71; Raab 27
37	Fall	111	6	Melotte 38; Collinder 75; Raab 28
38	Fall	87	2	Melotte 36; Collinder 67; Raab 26
39	Summer	79	5	Melotte 236; Collinder 438
40	Spring	46	2	
41	Winter	232	4	Melotte 52; Collinder 118; Raab 40
42	Fall	207	3	
43	Fall	207	3	
44	Winter	138	5	Melotte 88; Collinder 189; Raab 75; Beehive Cluster; Praesepe
45	Fall	133	3A	Melotte 22; Collinder 42
46	Winter	209	9	Melotte 75; Collinder 159; Raab 62
47	Winter	209	6	Melotte 68; Collinder 152; Raab 55
48	Winter	210	1	Melotte 85; Collinder 179; Raab 72
49	Spring	166	6	
50	Winter	209	1	Melotte 58; Collinder 124; Raab 45
51	Spring	71	4	Whirlpool Galaxy; 4C 47; UGC 8493; Arp 85
52	Summer	57	5	Melotte 243; Collinder 455; Raab150
53	Spring	143	2	
54	Summer	268	5	
55	Summer	269	5	
56	Summer	125	2	
57	Summer	124	5	Ring Nebula; PK 63+13.1
58	Spring	166	5	
59	Spring	166	8	
60	Spring	166	8	
61	Spring	190	1	
62	Spring	267	2	
63	Spring	95	1	Sunflower Galaxy
64	Spring	142	3	Black Eye Galaxy
65	Winter	165	1	Arp 317
66	Winter	165	1	
67	Winter	162	3	Melotte 94; Collinder 204; Raab 81
68	Spring	262	4	
69	Summer	268	5	

Table A.6. (*Continued*)

Messier No.	Season	Zone	Map	Other Names
70	Summer	268	5	
71	Summer	149	8	Melotte 226; Collinder 409
72	Summer	222	4	
73	Summer	222	4	Collinder 426
74	Fall	131	4	
75	Summer	246	2	
76	Fall	59	3	Little Dumbbell; Butterfly Nebula
77	Fall	180	4	
78	Fall	183	8	
79	Fall	231	2	
80	Spring	242	2	
81	Winter	19	2	
82	Winter	19	2	
83	Spring	263	4	
84	Spring	166	4	
85	Spring	142	2	
86	Spring	166	4	
87	Spring	166	4	Arp 152
88	Spring	166	4	
89	Spring	166	4	
90	Spring	166	4	Arp 76
91	Spring	166	4	
92	Spring	99	1	
93	Winter	233	9	Melotte 76; Collinder 160; Raab 63
94	Spring	94	3	
95	Winter	164	4	
96	Winter	164	6	
97	Winter	45	2	Owl Nebula
98	Spring	166	1	
99	Spring	166	1	Pinwheel Nebula
100	Spring	142	2	
101	Spring	72	1	Pinwheel Galaxy
102	Spring	49	2	
103	Fall	35	13	Melotte 8; Collinder 14; Raab 4
104	Spring	214	4	Sombrero Galaxy
105	Winter	164	5	
106	Spring	70	2	
107	Spring	218	4	
108	Winter	45	2	
109	Winter	69	3	
110	Fall	82	3	

Here a color-coded matrix indicates object density by the zones.

Double Stars Density

Dec	00	01	02	03	04	05	06	07	08	09	10	11	12	13	14	15	16	17	18	19	20	21	22	23
+85-90	Zone 1																							
+75-85	2			3			4			5			6			7			8			9		
+65-75	10	11	12	13	14	15	16	17	18	19	20	21	22	23	24	25	26	27	28	29	30	31	32	33
+55-65	34	35	36	37	38	39	40	41	42	43	44	45	46	47	48	49	50	51	52	53	54	55	56	57
+45-55	58	59	60	61	62	63	64	65	66	67	68	69	70	71	72	73	74	75	76	77	78	79	80	81
+35-45	82	83	84	85	86	87	88	89	90	91	92	93	94	95	96	97	98	99	100	101	102	103	104	105
+25-35	106	107	108	109	110	111	112	113	114	115	116	117	118	119	120	121	122	123	124	125	126	127	128	129
+15-25	130	131	132	133	134	135	136	137	138	139	140	141	142	143	144	145	146	147	148	149	150	151	152	153
+5-15	154	155	156	157	158	159	160	161	162	163	164	165	166	167	168	169	170	171	172	173	174	175	176	177
-5-+5	178	179	180	181	182	183	184	185	186	187	188	189	190	191	192	193	194	195	196	197	198	199	200	201
-5-15	202	203	204	205	206	207	208	209	210	211	212	213	214	215	216	217	218	219	220	221	222	223	224	225
-15--25	226	227	228	229	230	231	232	233	234	235	236	237	238	239	240	241	242	243	244	245	246	247	248	249
-25-35	250	251	252	253	254	255	256	257	258	259	260	261	262	263	264	265	266	267	268	269	270	271	272	273
Hour Angle	00	01	02	03	04	05	06	07	08	09	10	11	12	13	14	15	16	17	18	19	20	21	22	23

Deep Sky Density

Dec	00	01	02	03	04	05	06	07	08	09	10	11	12	13	14	15	16	17	18	19	20	21	22	23
+85-90	Zone 1																							
+75-85	2			3			4			5			6			7			8			9		
+65-75	10	11	12	13	14	15	16	17	18	19	20	21	22	23	24	25	26	27	28	29	30	31	32	33
+55-65	34	35	36	37	38	39	40	41	42	43	44	45	46	47	48	49	50	51	52	53	54	55	56	57
+45-55	58	59	60	61	62	63	64	65	66	67	68	69	70	71	72	73	74	75	76	77	78	79	80	81
+35-45	82	83	84	85	86	87	88	89	90	91	92	93	94	95	96	97	98	99	100	101	102	103	104	105
+25-35	106	107	108	109	110	111	112	113	114	115	116	117	118	119	120	121	122	123	124	125	126	127	128	129
+15-25	130	131	132	133	134	135	136	137	138	139	140	141	142	143	144	145	146	147	148	149	150	151	152	153
+5-15	154	155	156	157	158	159	160	161	162	163	164	165	166	167	168	169	170	171	172	173	174	175	176	177
-5-+5	178	179	180	181	182	183	184	185	186	187	188	189	190	191	192	193	194	195	196	197	198	199	200	201
-5-15	202	203	204	205	206	207	208	209	210	211	212	213	214	215	216	217	218	219	220	221	222	223	224	225
-15--25	226	227	228	229	230	231	232	233	234	235	236	237	238	239	240	241	242	243	244	245	246	247	248	249
-25-35	250	251	252	253	254	255	256	257	258	259	260	261	262	263	264	265	266	267	268	269	270	271	272	273
Hour Angle	00	01	02	03	04	05	06	07	08	09	10	11	12	13	14	15	16	17	18	19	20	21	22	23

Sample Zone

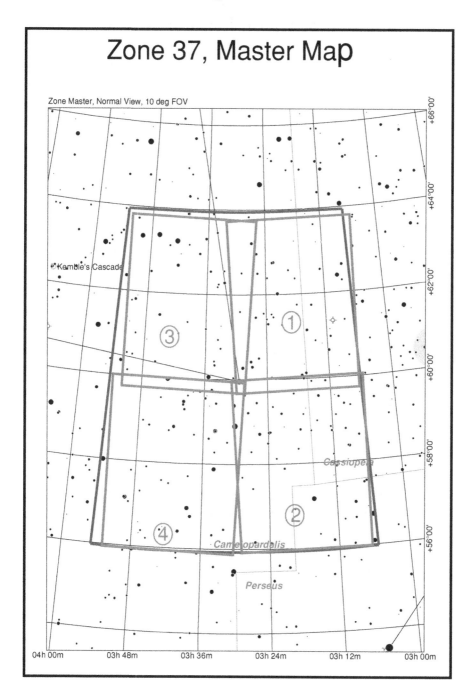

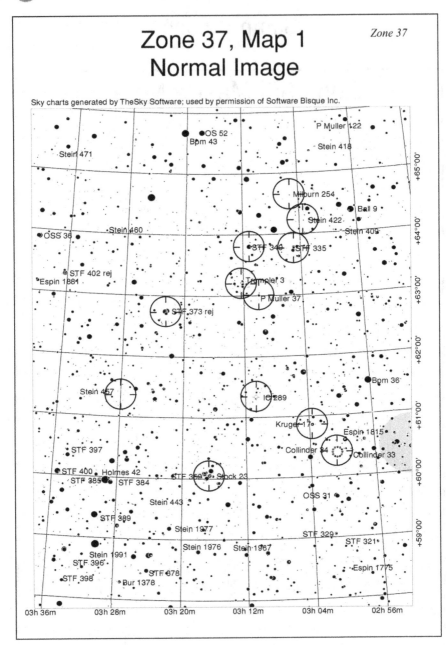

Zone 37, Map 1
Normal Image

Zone 37

Sky charts generated by TheSky Software; used by permission of Software Bisque Inc.

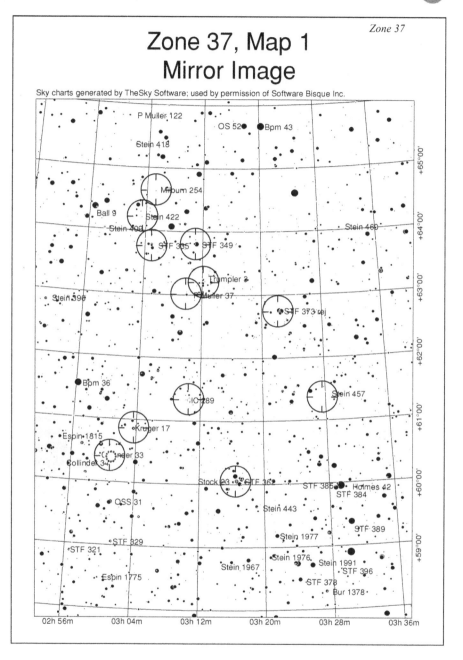

Zone 37

Zone 37, Map 1
Mirror Image

Sky charts generated by TheSky Software; used by permission of Software Bisque Inc.

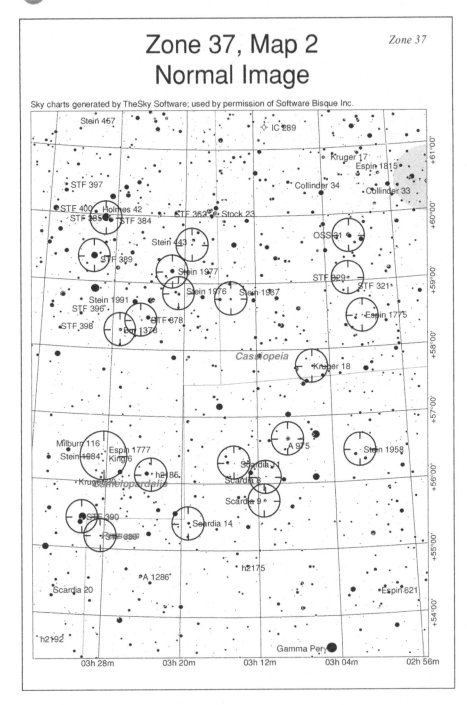

Zone 37, Map 2
Normal Image

Zone 37

Sky charts generated by TheSky Software; used by permission of Software Bisque Inc.

Stein 457
IC 289
Kruger 17
Espin 1815
Collinder 34
Collinder 33
STF 397
STF 400 Holmes 42
STF 385 STF 384
STF 362 Stock 23
Stein 443
OSS 31
STF 389
Stein 1977
STF 329
STF 321
Stein 1976 Stein 1987
Stein 1991
STF 396
Espin 1775
STF 398
STF 878
Bu 1378
Cassiopeia
Kruger 18
Milburn 116
Espin 1777
A 975
Stein 1958
Stein 1984 King 6
h 2186
Scardia 11
Kruger *Camelopardalis*
Scardia 8
Scardia 9
STF 390
Scardia 14
STF 686
A 1286
h 2175
Scardia 20
Espin 621
h 2192
Gamma Per

03h 28m　03h 20m　03h 12m　03h 04m　02h 56m

+61°00'
+60°00'
+59°00'
+58°00'
+57°00'
+56°00'
+55°00'
+54°00'

Zone 37, Map 2
Mirror Image

Zone 37

Sky charts generated by TheSky Software; used by permission of Software Bisque Inc.

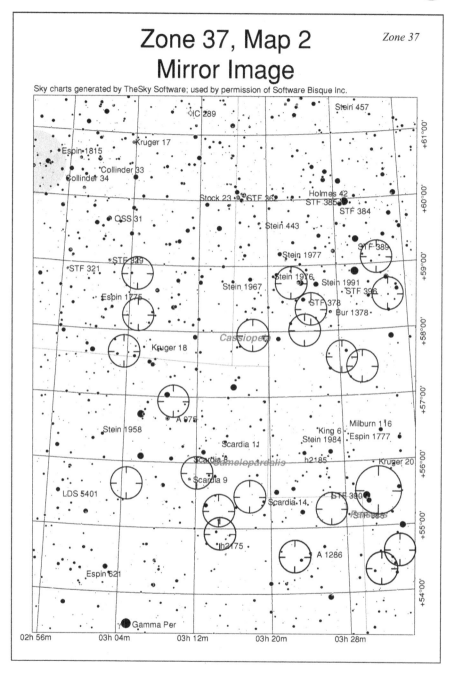

IC 289
Stein 457
Kruger 17
Espin 1815
Collinder 33
Collinder 34
Stock 23 · STF 362
Holmes 42
STF 385
STF 384
QSS 31
Stein 443
STF 389
Stein 1977
STF 349
Stein 1976
STF 321
Stein 1967
Stein 1991
STF 396
Espin 1775
STF 378
Bur 1378
Cassiopeia
Kruger 18
A 975
Milburn 116
Stein 1958
King 6
Stein 1984
Espin 1777
Scardia 11
Camelopardalis
h2185
Kruger 20
Scardia 8
Scardia 9
LDS 5401
STF 390
Scardia 14
STF 393
h2175
A 1286
Espin 621
Gamma Per

02h 56m 03h 04m 03h 12m 03h 20m 03h 28m

+61°00' +60°00' +59°00' +58°00' +57°00' +56°00' +55°00' +54°00'

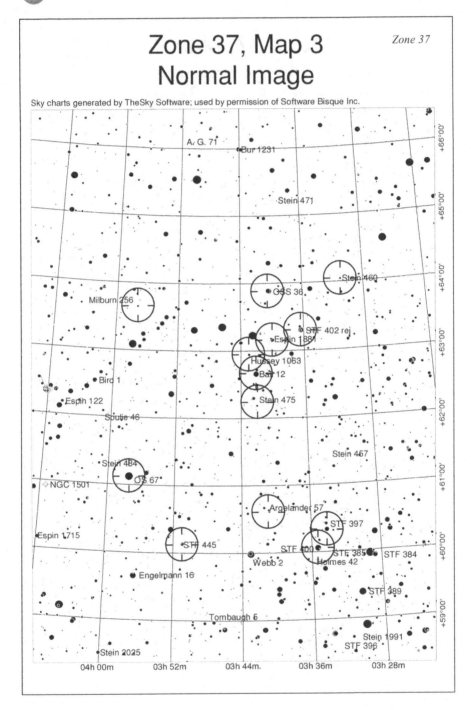

Zone 37, Map 3
Normal Image

Zone 37

Sky charts generated by TheSky Software; used by permission of Software Bisque Inc.

A. G. 71
Bur 1231
Stein 471
Stein 460
GSS 36
Milburn 256
STF 402 rej
Espin 1881
Hussey 1063
Bar 12
Bird 1
Stein 475
Espin 122
Soutie 46
Stein 467
Stein 484
OS 67
NGC 1501
Argelander 57
Espin 1715
STF 397
STF 445
STF 400
Webb 2
STF 385
Holmes 42
STF 384
Engelmann 16
STF 389
Tombaugh 5
Stein 1991
STF 396
Stein 2025

04h 00m   03h 52m   03h 44m.   03h 36m   03h 28m

+66°00'
+65°00'
+64°00'
+63°00'
+62°00'
+61°00'
+60°00'
+59°00'

Zone 37, Map 3
Mirror Image

Zone 37

Sky charts generated by TheSky Software; used by permission of Software Bisque Inc.

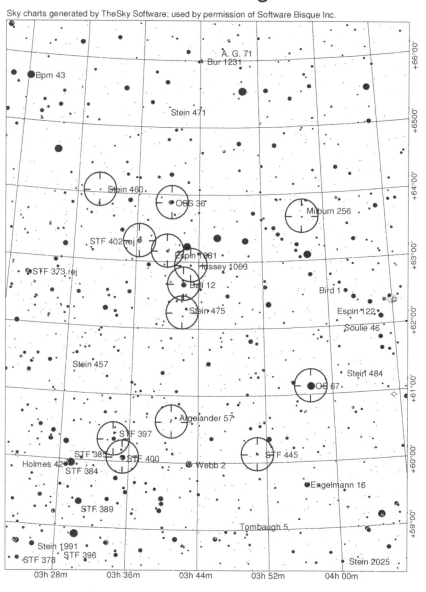

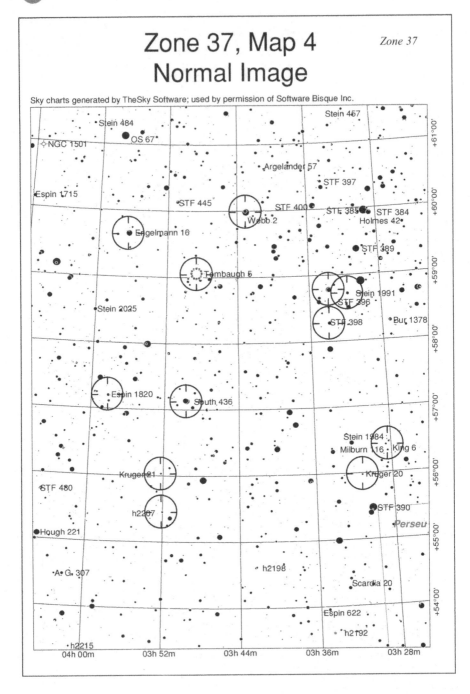

Zone 37, Map 4
Normal Image

Zone 37

Sky charts generated by TheSky Software; used by permission of Software Bisque Inc.

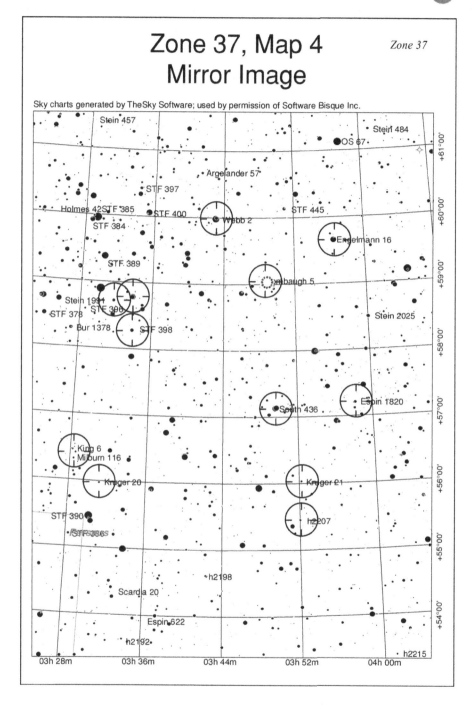

Zone 37, Map 4
Mirror Image

Zone 37

Sky charts generated by TheSky Software; used by permission of Software Bisque Inc.

Stein 457

Stein 484

OS 67

Argelander 57

STF 397

Holmes 42 STF 385   STF 400   STF 445

STF 384   Webb 2   Engelmann 16

STF 389

baugh 5

Stein 1991   STF 378   STF 39   Stein 2025

Bur 1378   STF 398

Espin 1820

Smith 436

King 6
Milburn 116

Kruger 20   Kruger 21

STF 390   h2207

STF 386

h2198

Scardia 20

Espin 622

h2192   h2215

+61°00'
+60°00'
+59°00'
+58°00'
+57°00'
+56°00'
+55°00'
+54°00'

03h 28m   03h 36m   03h 44m   03h 52m   04h 00m

Zone 37

Zone 37, Map 1

| Double Stars |

Easy

| STF 335 | Rating: 3 E |

Other name(s): **HD 18800; SAO 12589**

Position: 0305+6345

	Magnitude	Separation	PA	Year	Spectra	Colors
A	8.61	—	—	—	F5 V	W
B	9.46	21.9 -	161 +	1991		O

Notes:
   1831: 24.4 @ 159. Sixteen measurements. Hipparcos/Tycho data show different distances for these stars; they may be optical. The stars show a similar, but small, proper motion.

Observations:
   C-8 at 104x. It is the obtuse angle of an 8.5-m triangle.

Scale model (in which the Sun is the size of a baseball):
         Diameter: A = 12.0 inches.
         Separation: AB = 12.4 miles.

Distance (LY): 362   Total luminosity (Suns): 6

| STF 349 | Rating: 2 E |

Other name(s): **ADS 2371; HD 19440; SAO 12635**

Position: 0311+6347

	Magnitude	Separation	PA	Year	Spectra	Colors
A	7.43	—	—	—	F8 V	W
B	8.14	6.1 =	317 -	1994		Y

Notes:
   1832: 6.1 @ 320. Fourty-two measurements. The stars show a similar, but large, proper motion.

Observations:
    C-8 at 104x. There is a delightful and very rich field about 50' S of this star. It looks like a good Messier open cluster!

Scale model (in which the Sun is the size of a baseball):
    Diameter: A = 11.1 inches.
    Separation: AB = 2.48 miles.

Distance (LY): 260   Total luminosity (Suns): 7.7

STF 362					**Rating: 2 E**	

Other name(s): **ADS 2426; HD 20053; SAO 23908**

Position: 0316+6002

	Magnitude	Separation	PA	Year	Spectra	Colors
A	8.30	—	—	—	A0 V	W
B	8.80	7.3 +	137 -	1994		W
C	10.50	26.7 +	46 +	1987		?
D	11.10	29.9 -	287 +	1987		No
E	9.90	35.3 =	243 +	1987		No
F	11.00	106.6	258	1987		No
G	13.72	110.6	281	1987		No

Notes:
    AB 1831: 7.1 @ 143. Fourty-five measurements. The stars have a small difference in proper motion.
    AC 1893: 26.0 @ 42. Four measurements.
    AD 1915: 30.9 @ 286. Two measurements.
    AE 1866: 35.3 @ 242. Seven measurements.
    AF and AG, One measurement.

Observations:
    C-8 at 83x. Beautiful! The field is Stock 23. Bright and rich.

Scale model (in which the Sun is the size of a baseball):
    Diameter: A = 5.76 inches.
    Separation: AB = 2.06 miles; AC = 7.54 miles; AD = 8.45 miles; AE = 9.98 miles; AF = 30.1 miles; AG = 31.2 miles.

Distance (LY): 181   Total luminosity (Suns): 1.7

STF 373 rej	Rating: 2 E

Other name(s): **OS 33 rej (C); HD 20588; SAO 12721**

Position: **0322+6244**

	Magnitude	Separation	PA	Year	Spectra	Colors
A	7.63	—	—	—	F8 IV	W
B	10.40	20.1 +	118 +	1991	G	B
C	7.78	116.2 -	111 +	1991	A0	bW
D		179.8	166	1984		No

Notes:
AB 1875: 19.8 @ 117. Eleven measurements. Hipparcos/Tycho data show different distances for these stars; they may be optical. However, the stars exhibit common proper motion.
AC 1875: 117.7 @ 110. Fifteen measurements.
AD 1 measurement.

Observations:
C-8 at 83x. Nice! There is a 9.0-m star 4' S (SAO 12722) that makes for a nice, bright triangle pointing S. An 11-m star is 20" away in pa 110; and an 8.8-m star is 115" in pa 175.

Scale model (in which the Sun is the size of a baseball):
Diameter: A = 11.3 inches; B = 13.2 inches.
Separation: AB = 16.3 miles; AC = 94.3 miles; AD = 146 miles.

Distance (LY): 520   Total luminosity (Suns): 41

Moderate

Stein 422	Rating: 5 M

Position: **0303+6412**

	Magnitude	Separation	PA	Year	Spectra	Colors
A	10.14	—	—	—		W
B	10.71	9.8 -	126 -	1906		W

Notes:
1904: 10.1 @ 127. Four measurements.

Observations:
C-11 at 115x.

Zone 37

Distance (LY): 408   Total luminosity (Suns): 1.82

Kruger 17					**Rating: 4 M**	

Other name(s): **ADS 2299**

Position: 0303+6051

	Magnitude	Separation	PA	Year	Spectra	Colors
A	8.77	—	—	—	F7 V	W
B	9.45	3.2 -	219 -	1991		Y

Notes:
1890: 3.5 @ 221. Sixteen measurements.

Observations:
C-8 at 280x. It is the S corner of a 9m "kite" of stars whose long axis runs NE/SW. Rich, bright field.

Scale model (in which the Sun is the size of a baseball):
Diameter: A = 7.50 inches.
Separation: AB = 1.90 miles.

Distance (LY): 375   Total luminosity (Suns): 4

Milburn 254					**Rating: 5 M**	

Other name(s): **ADS 2317**

Position: 0305+6438

	Magnitude	Separation	PA	Year	Spectra	Colors
A	10.68	—	—	—		W
B	11.90	6.6 +	192 +	1922		?

Notes:
1906: 6.3 @ 185. Two measurements.

Observations:
C11 at 115x.

P Muller 37					Rating: 5 M	

Other name(s): **SAO 12624**

Position: 0310+6259

	Magnitude	Separation	PA	Year	Spectra	Colors
A	9.65	—	—	—	B5	W
B	11.00	0.5 =	207 +	1996		??
C	12.18	17.0	332			?

Notes:
AB 1969: 0.5 @ 198. Five measurements.
AC 1969: 17.0 @ 332. One measurement.

Observations:
C-11 at 115x. B was not seen at any magnifcation.

Scale model (in which the Sun is the size of a baseball):
Diameter: A = 2.99 inches.
Separation: AB = 1.59 miles; AC = 54.1 miles.

Distance (LY): 2040   Total luminosity (Suns): 62.2

Stein 457					Rating: 5 M	

Position: 0328+6124

	Magnitude	Separation	PA	Year	Spectra	Colors
A	11.17	—	—	—		W
B	11.52	13.9 -	157 -	1991		bW

Notes:
1904: 15.8 @ 341. Seven measurements. The 1991 measurement is a quadrant reversal.
The Digitized Sky Survey 2 shows 13.1 @ 331.

Observations:
C-11 at 115x.

Distance (LY): 19   Total luminosity (Suns): 0.002

Zone 37

Deep Sky Objects

Easy

| **Stock 23** | **Rating: 3 E** |

Position: 0316+6002

Type : Oc/Gn   Class: III 3pn   Magnitude : 6.5
Dimensions: 15'   Cluster population: 25

Observations:
    C-8 at 83x. Very bright. Dominated by a keystone (the SW member of which is a close double). I counted 18 stars at low power.

Moderate

| **Collinder 34** | **Rating: 5 M** |

Position: 0301+6025

Type : Oc   Class: I 3p   Magnitude : 6.8
Dimensions: 25'

Notes:
    This cluster is actually in front of Collinder 33.

Observations:
    C-8 at 104x. Not much more than a bright richening of the moderate and bright field.

| **Trumpler 3** | **Rating: 4 M** |

Other Names: Collinder 36, Harvard 1

Position: 0312+6315

Type : Oc   Class: III 3p   Magnitude : 7.0
Dimensions: 23.0'   Cluster population: 30

Observations:
    C-8 at 104x. There are richer fields in the Milky Way! Steve Coe aptly writes, "this is one of the best Trumpler or Harvard clusters I remember seeing." That is because most of the Trumpler and Harvard clusters are poor specimens in small telescopes!

Zone 37

Difficult

IC 289	Rating: 5 D

Other Names: PK 138+2.1

Position: 0310+6119

Type : Pn   Class: 4   Magnitude : 12.3
Dimensions: 45" x 30"

Notes:
    Discovered by Swift in 1888, it lies just to the south of the 10th magnitude star BD
+ 60 0631, which makes it a fairly easy find. At high magnification it appears to "flip"
orientation!
    IC 289 has been described as faint and homogeneous in an 8-inch. A UHC or O-III
filter helps make it more visible. The UHC is recommended for scopes less than 14-
inches, but the OIII will show more detail in larger instruments. N. J. Martin described IC
289 as, "a nice, faint round planet like planetary nebula. The uniform oval disc shows
some irregularity in brightness but is not obviously brighter at the edge."
    (http://www.skyhound.com/sh/archive/oct/NGC_289.html)

Observations:
    C-8 at 83x. It is very faint and in a line with 10–12 m stars.
    C11 at 115x with O-III filter. Even with the filter, it was very faint, large and of low
surface brightness. No details to report.

Model (where the Sun is a baseball):
    Its diameter would be 274 miles.

    Distance (LY): 3900   Luminosity (suns): 14.0

Zone 37, Map 2

Double Stars

Easy

Espin 1775	Rating: 3 E

Other name(s): **ADS 2259**

Position: 0300+5828

		Magnitude	Separation	PA	Year	Spectra	Colors
	A	10.24	—	—	—		W
	B	10.60	2.0 -!	288 +	1992		W

Notes:
　1908: 7.1 @ 283. Seven measurements.

Observations:
　C-11 at 339x.

　Distance (LY): 267 　　　Total luminosity (Suns): 0.768

OSS 31	Rating: 4 E

Other name(s): **HD 18473; SAO 23761**

Position: 0301+5940

		Magnitude	Separation	PA	Year	Spectra	Colors
	A	7.35	—	—	—	B9p	W
	B	8.07	73.4 -	230 +	1999	A0	pgW

Notes:
　1875: 73.6 @ 229. Nineteen measurements.
　Star A is a spectroscopic binary.
　The stars share common proper motion.

Observations:
    C-8 at 83x. Moderate field.

Scale model (in which the Sun is the size of a baseball):
        Diameter: A = 4.82 inches; B = 6.15 inches.
        Separation: AB = 84.8 miles.

    Distance (LY): 740   Total luminosity (Suns): 75

STF 329					**Rating: 4 E**	

Other name(s): **ADS 2269; HD 18518; SAO 23764**

Position: 0301+5902

	Magnitude	Separation	PA	Year	Spectra	Colors
A	8.66	—	—	—	A2 IV	yW
B	10.36	16.1 +	274 +	1991		pB

Notes:
    1830: 16.0 @ 272. Sixteen measurements. Hipparcos/Tycho data show different distances for these stars; they may be optical. The stars show a similar, but small, proper motion.

Observations:
    C8 at 83x.

Scale model (in which the Sun is the size of a baseball):
        Diameter: A = 5.90 inches.
        Separation: AB = 75.4 miles.

    Distance (LY): 3000   Total luminosity (Suns): 380

Zone 37

Stein 1958					Rating: 4 E

Position: 0301+5627

	Magnitude	Separation	PA	Year	Spectra	Colors
A	10.54	—	—	—		W
B	11.72	14.3 +	133 -	1991		?

Notes:
   1908: 13.6 @ 135. Five measurements.

Observations:
   C-11 at 115x. It is the right angle of a 30/60/90 right triangle with two 9-m stars.

Scardia 8					Rating: 5 E

Position: 0311+5605

	Magnitude	Separation	PA	Year	Spectra	Colors
A	10.02	—	—	—		W
B	11.38	23.5 +	176 -	1991		W

Notes:
   1911: 23.4 @ 178. Four measurements.
   Hipparcos/Tycho data show different distances for these stars (75 and 205 light years); they may be optical.

Observations:
   C-8 at 104x. It gets lost in a nice, faint but rich field.

Scardia 9	Rating: 3 E

Other name(s): **HD 19537; SAO 23856**

Position: **0311+5544**

	Magnitude	Separation	PA	Year	Spectra	Colors
A	8.73	—	—	—		W
B	10.36	27.8 -	137 =	1991		bW
C	11.32	53.7 +	239 +	1987		?

Notes:
　　AB 1911: 28.0 @ 137. Five measurements. No evidence of duplicity by seen by Hipparcos.
　　AC 1911: 50.5 @ 227. Four measurements.
　　A appears to be 212 light years away; B 145.

Observations:
　　C-11 at 115x.

Scardia 11	Rating: 5 E

Position: **0314+5617**

	Magnitude	Separation	PA	Year	Spectra	Colors
A	10.74	—	—	—		W
B	12.31	26.2	267	1985		?

Notes:
　　One measurement.

Observations:
　　C-11 at 115x.

Appendix

Stein 443					Rating: 5 E

Position: 0319+5937

	Magnitude	Separation	PA	Year	Spectra	Colors
A	10.72	—	—	—		W
B	11.8	10.6	77	1908		W

Notes:
One measurement.

Observations:
C-11 at 115x. An 8-m star that lies 1' N is a good skymark.

Distance (LY): 135   Total luminosity (Suns): 0.101

Scardia 14					Rating: 5 E

Other name(s): **HD 20364; SAO 23945**

Position: 0319+5524

	Magnitude	Separation	PA	Year	Spectra	Colors
A	9.00	—	—	—		W
B	10.49	46.9 +	293 =	1991		bW

Notes:
1911: 46.6 @ 293. Seven measurements.

Observations:
C-8 at 104x.

Stein 1976   Rating: 4 E

Other name(s): **HD 237119; SAO 23956**

Position: **0320+5850**

	Magnitude	Separation	PA	Year	Spectra	Colors
A	9.34	—	—	—		W
B	12.13	13.8 -	63 -	1920		?

Notes:
     1908: 14.2 @ 66. Three measurements.

Observations:
     C-11 at 115x.

---

h2185   Rating: 4 E

Position: **0323+5607**

	Magnitude	Separation	PA	Year	Spectra	Colors
A	10.55	—	—	—		W
B	11.60	9.3 -	249 +	1911		bW

Notes:
     1909: 9.5 @ 248. Two measurements.

Observations:
     C-11 at 115x.

STF 378   Rating: 5 E

Other name(s): **ADS 2505; HD 20847; SAO 24008**

Position: **0324+5826**

	Magnitude	Separation	PA	Year	Spectra	Colors
A	9.55	—	—	—	F	W
B	10.64	18.7 +	314 +	1991		W

Notes:
     1830: 18.6 @ 313. Fourteen measurements.
     The stars show a similar, but small, proper motion.

Observations:
    C-8 at 104x. Overpowered by the rich, bright field.

Scale model (in which the Sun is the size of a baseball):
    Diameter: A = 14.2 inches.
    Separation: AB = 6.65 miles.

Distance (LY): 228   Total luminosity (Suns): 0.8

Bur 1378				Rating: 4 E	

Other name(s): **HD 21084; SAO 24034**

Position: 0327+5816

	Magnitude	Separation	PA	Year	Spectra	Colors
A	8.52	—	—	—	F8	W
B	12.39	26.4 -	124 +	1920		?

Notes:
    AB 1904: 26.7 @ 122. Four measurements.

Observations:
    C-11 at 115x.

Scale model (in which the Sun is the size of a baseball):
    Diameter: A = 15.4 inches.
    Separation: AB = 6.38 miles.

Distance (LY): 155   Total luminosity (Suns): 0.755

Stein 1984				Rating: 4 E	

Position: 0328+5627

	Magnitude	Separation	PA	Year	Spectra	Colors
A	11.23	—	—	—		W
B	12.07	11.1 -	74 -	1991		?

Notes:
    1911: 11.4 @ 77. Four measurements.

Zone 37

Observations:
    C-11 at 115x.

Distance (LY): 26   Total luminosity (Suns): 0.003

Espin 1777	Rating: 4 E

Other name(s): **ADS 2539**

Position: 0328+5628

	Magnitude	Separation	PA	Year	Spectra	Colors
A	10.79	—	—	—		W
B	10.90	2.3 =	245 +	1983		W?

Notes:
    1919: 2.3 @ 244. Two measurements.

Observations:
    C-11 at 339x. Nice field.

STF 390	Rating: 4 E

Other name(s): **ADS 2565; HD 21447; SAO 24064**

Position: 0330+5527

	Magnitude	Separation	PA	Year	Spectra	Colors
A	5.06	—	—	—	A1 V	W
B	10.02	14.6 -	160 =	1999		B
C	10.63	110.2 +	171 -	1991		R

Notes:
    AB 1832: 15.0 @ 160. Twelve measurements.
    AC 1910: 110.1 @ 172. Seven measurements.
    Star A rotates at 179 kps and is a possible member of the Eggen group, a subset of
    the Ursa Major moving stream. It is also a Delta Scuti variable.
    The AB stars show similar proper motions.

Observations:
    C-8 at 83x. Some observers report Y and V. Webb saw gW. Franks saw yW and
pB.

Zone 37

Scale model (in which the Sun is the size of a baseball):
      Diameter: A = 8.70 inches.
      Separation: AB = 4.35 miles; AC = 32.8 miles.

Distance (LY): 191   Total luminosity (Suns): 27

Moderate

Stein 1967	Rating: 5 M

Position: 0314+5846

	Magnitude	Separation	PA	Year	Spectra	Colors
A	9.78	—	—	—		yW
B	11.90	9.3 =	95 +	1920		W

Notes:
    1908: 9.3 @ 90. Four measurements.

Observations:
    C-11 at 115x.

Stein 1977	Rating: 4 M

Other name(s): **HD 20547**

Position: 0321+5910

	Magnitude	Separation	PA	Year	Spectra	Colors
A	8.20	—	—	—	B8	W
B	10.16	5.4 +	66 -	1993		rO

Notes:
    1908: 4.7 @ 74. Six measurements.

Observations:
    C-8 at 104x.

Scale model (in which the Sun is the size of a baseball):
      Diameter: A = 4.32 inches.
      Separation: AB = 13.5 miles.

Distance (LY): 1600   Total luminosity (Suns): 111

STF 386					**Rating: 4 M**	

Other name(s): **ADS 2537; HD 21225**

Position: 0328+5510

	Magnitude	Separation	PA	Year	Spectra	Colors
A	9.46	—	—	—	A0	W
B	9.40	2.6 +	60 +	1998		Y

Notes:
> 1830: 2.5 @ 59. Fourteen measurements.

Observations:
> C-8 at 280x. Faint field.

Milburn 116					**Rating: 5 M**	

Other name(s): **ADS 2536**

Position: 0328+5625

	Magnitude	Separation	PA	Year	Spectra	Colors
A	10.34	—	—	—		W
B	10.60	3.3 +	47 +	1978		?
C	10.60	13.8 +	114 -	1919		?

Notes:
> AB 1911: 3.2 @ 45. Three measurements.
> AC 1911: 13.7 @ 129. Two measurements.

Observations:
> C-11 at 115x.

STF 389					**Rating: 3 M**	

Other name(s): **ADS 2563; HD 21427; SAO 24062**

Position: 0330+5922

	Magnitude	Separation	PA	Year	Spectra	Colors
A	6.14	—	—	—	A2 V	W
B	7.50	2.5 -	71 +	1997		W

Notes:
> 1831: 2.8 @ 62. Seventy-one measurements.
> McAlister 11 is a star at 0.3 @ 61 (1989).

Observations:
> C-8 at 206x. Webb saw them as W, V. Moderate field.-

Scale model (in which the Sun is the size of a baseball):
> Diameter: A = 5.00 inches.
> Separation: AB = 1.31 miles.

Distance (LY): 337   Total luminosity (Suns): 42

Holmes 42					Rating: 5 M	

Other name(s): **ADS 2557; HD 237146; SAO 24060**

Position: 0330+5955

	Magnitude	Separation	PA	Year	Spectra	Colors
A	9.40	—	—	—	B3	W
B	10.90	5.3 =	52 +	1992		W?

Notes:
> 1901: 5.3 @ 51. Fourteen measurements.

Observations:
> C-11 at 115x. A 4.6-m star 1' W cooks it! (This is STF 385.)

Difficult

Kruger 18					Rating: 4 D	

Other name(s): **ADS 2327; SAO 23792**

Position: 0306+5744

	Magnitude	Separation	PA	Year	Spectra	Colors
A	9.83	—	—	—	A2	W
B	10.17	1.5 +	273 -	2000		W

Notes:
> 1890: 1.2 @ 274. Fourteen measurements.

Observations:
> C-11 at 634x.

A 975					Rating: 3 D	

Other name(s): **ADS 2360; HD 19278; SAO 23830**

Position: **0309+5639**

	Magnitude	Separation	PA	Year	Spectra	Colors
A	8.44	—	—	—	K0	O
B	10.15	1.6 =	203 +	1991		yW

Notes:
    1905: 1.6 @ 202. Eight measurements.

Observations:
    C-11 at 339x.

STF 384					Rating: 5 D	

Other name(s): **ADS 2540**

Position: **0328+5954**

	Magnitude	Separation	PA	Year	Spectra	Colors
A	8.16	—	—	—	F8	W
B	8.96	2.0 =	274 +	1998		W
C	10.40	116.8 -	340 =	1908		No

Notes:
    AB 1830: 2.0 @ 268. Twenty-seven measurements.
    AC 1879: 116.9 @ 340. Two measurements.

Observations:
    C-8 at 280x. Webb said he saw D and dB.

Scale model (in which the Sun is the size of a baseball):
    Diameter: A = 15.1 inches.
    Separation: AB = 11.1 miles; AC = 647 miles (probably optical).

Distance (LY): 3550   Total luminosity (Suns): 1230

Appendix

Zone 37

STF 385	Rating: 4 D

Other name(s): **ADS 2544; HD 21291; SAO 24054**

Position: 0329+5956

	Magnitude	Separation	PA	Year	Spectra	Colors
A	4.33	—	—	—	B9 Ia	Y
B	7.97	2.5 +	159 -	1991		O

Notes:
    1829: 2.4 @ 161. Twenty-one measurements. It is a member of the Cam OB1 Association.

Observations:
    C-8 at 206x. Difficult.

Scale model (in which the Sun is the size of a baseball):
        Diameter: A = 10.2 inches.
        Separation: AB = 16.8 miles.

    Distance (LY): 4300   Total luminosity (Suns): 31,900

Zone 37, Map 3

| Double Stars |

Easy

| STF 397 | | | | Rating: 5 E |

Other name(s): **ADS 2598; HD 21801**

Position: 0334+6024

	Magnitude	Separation	PA	Year	Spectra	Colors
A	8.61	—	—	—	A0 Ia?	W
B	9.10	5.0 -	41 -	1993		B

Notes:
   1829: 5.1 @ 43. Ten measurements.

Observations:
   C-11 at 115x.

Scale model (in which the Sun is the size of a baseball):
   Diameter: A = 5.86 inches.
   Separation: AB = 254 miles.

   Distance (LY): 32,600   Total luminosity (Suns): 49,000

| STF 402 rej | | | | Rating: 4 E |

Other name(s): **ADS 2617; HD 21971; SAO 12830**

Position: 0336+6317

	Magnitude	Separation	PA	Year	Spectra	Colors
A	7.30	—	—	—	K4 III	yO
B	10.40	12.0 -	164 +	1962		W?

Notes:
   1902: 12.5 @ 161. Four measurements.
   Hipparcos/Tycho data show different distances for these stars; they may be optical.

Observations:
   C-8 at 206x. Rich field.

Zone 37

Scale model (in which the Sun is the size of a baseball):
    Diameter: A = 34.0 inches.
    Separation: AB = 46.8 miles.

Distance (LY): 2500   Total luminosity (Suns): 634

OSS 36				Rating: 4 E	

Other name(s): **Hussey 1062 (C); ADS 2650**

Position: 0340+6352

	Magnitude	Separation	PA	Year	Spectra	Colors
A	6.81	—	—	—	F5 V	W
B	8.21	46.2 +	71 +	1993		Y

Notes:
    1875: 45.8 @ 70. Nineteen measurements. The stars share common proper motion.

Observations:
    C-8 at 104x. Webb saw them as W and Y, Espin as yW and oW.
    C-11 at 115x.

Scale model (in which the Sun is the size of a baseball):
    Diameter: A = 10.6 inches.
    Separation: AB = 10.3 miles.

Distance (LY): 142   Total luminosity (Suns): 4.02

Ball 12				Rating: 5 E	

Other name(s): **U Cam; Hipparcos 470 (a); HD 22611**

Position: 0342+6239

	Magnitude	Separation	PA	Year	Spectra	Colors
A	6.99	—	—	—	C6.4 ev	rP!
B	7.10	0.2	79	1991		??
C	9.66	208 =	349 =	1910		W

Notes:
    AB 1 measurement.
    AC 1879: 208 @ 349. Two measurements.
    A is a semiregular variable.

Observations:
   C-11 at 115x. A is an intense color.
   B was too close to resolve.

Scale model (in which the Sun is the size of a baseball):
   Separation: AB = 1.28 miles; AC = 1,331 miles (probably optical).

Distance (LY): 4100   Total luminosity (Suns): 4180

Hussey 1063				**Rating: 5 E**	

Other name(s): **Fox; ADS 2685; SAO 12879**

Position: 0343+6256

	Magnitude	Separation	PA	Year	Spectra	Colors
A	9.52	—	—	—	F0	W
B	10.52	2.8 =	335 =	1991		No
C	11.57	49.7 -	122 -	1912		?

Notes:
   AB 1905: 2.8 @ 335. Eight measurements.
   AC 1904: 50.7 @ 123. Three measurements.

Observations:
   C-8 at 104x. A 5.5-m star is to the N.

Moderate

Stein 460				**Rating: 5 M**	

Position: 0331+6402

	Magnitude	Separation	PA	Year	Spectra	Colors
A	10.36	—	—	—		W
B	11.50	10.1	310	1906		bW

Notes:
   One measurement.

Observations:
   C-11 at 115x.

Zone 37

Distance (LY): 91   Total luminosity (Suns): 0.063

Espin 1881   **Rating: 3 M**

Other name(s): **ADS 2657; SAO 12852**

Position: 0340+6310

	Magnitude	Separation	PA	Year	Spectra	Colors
A	10.50	—	—	—		W
B	10.60	2.6 -	161 -?	1978		W

Notes:
   1921: 3.0 @ 336. Two measurements.

Observations:
   C-11 at 339x.

Distance (LY): 67   Total luminosity (Suns): 0.042

Argelander 57   **Rating: 4 M**

Other name(s): **ADS 2664; HD 22577; SAO 12865**

Position: 0341+6040

	Magnitude	Separation	PA	Year	Spectra	Colors
A	9.59	—	—	—	A2	W
B	9.69	6.3 +	311 -?	1994		W

Notes:
   1898: 5.4 @ 146. Fifteen measurements.
   The stars share common proper motion. Quadrant reversal?

Observations:
   C-8 at 104x. Moderately rich field.

Scale model (in which the Sun is the size of a baseball):
         Diameter: A = 7.73 inches.
         Separation: AB = 2.23 miles.

Distance (LY): 227   Total luminosity (Suns): 1.3

| **Stein 475** | | | | | **Rating: 5 M** |

Position: 0342+6216

	Magnitude	Separation	PA	Year	Spectra	Colors
A	10.64	—	—	—		W
B	10.55	9.0 +	92 -	1906		W

Notes:
  1898: 8.7 @ 96. Four measurements.
  Hipparcos gives discordant distances – 47 light years and 86.

Observations:
  C-11 at 115x. The separatioin and PA looked both to be greater than listed.

| **STF 445** | | | | | **Rating: 4 M** |

Other name(s): **ADS 2791; HD 23726; SAO 24253**

Position: 0351+6007

	Magnitude	Separation	PA	Year	Spectra	Colors
A	8.56	—	—	—	B5	W
B	9.52	3.0 =	259 +	1991		B

Notes:
  1831: 3.0 @ 253. Ten measurements.
Observations:
  C-8 at 280x.

Scale model (in which the Sun is the size of a baseball):
        Diameter: A = 2.81 inches.
        Separation: AB = 6.92 miles.

    Distance (LY): 1480   Total luminosity (Suns): 117

Zone 37

OS 67					Rating: 4 M

Other name(s): **ADS 2867; HD 24480; SAO 12968**

Position: **0357+6107**

	Magnitude	Separation	PA	Year	Spectra	Colors
A	4.99	—	—	—	K4 II	O
B	8.16	1.7 =	49 +	1991		Y

Notes:
   1843: 1.7 @ 44. Thirty-six measurements.
   The primary is an infra-red source.
Observations:
   C-8 at 280x. Notched. Some observers report Y and G. Webb saw them as D and
G.

Scale model (in which the Sun is the size of a baseball):
      Diameter: A = 54.0 *feet!*.
      Separation: AB = 4.90 miles.

   Distance (LY): 1850   Total luminosity (Suns): 2910

Difficult

STF 400					Rating: 4 D

Other name(s): **ADS 2612; HD 21903; SAO 24111**

Position: **0335+6002**

	Magnitude	Separation	PA	Year	Spectra	Colors
A	6.86	—	—	—	F5 V	W
B	7.99	1.5 =	265 -	2000	F9 V	?
C	10.66	92.6 +	236 -	1985		B

Notes:
   AB 1829: 1.5 @ 283. Over 100 measurements. The orbit takes 287.7 years (Baize,
1952) with semimajor axis of 1.24" and direct motion. The combined mass of stars A
and B is 2.2 solar masses.
   AC 1908: 92.2 @ 238. Five measurements.
Observations:
   C-8 at 83x. Webb saw them as Y, bW. Very rich field.

Zone 37

Scale model (in which the Sun is the size of a baseball):
    Diameter: A = 10.7 inches; B = 11.5 inches.
    Separation: AB = 2,270 feet; AC = 26.5 miles.

Distance (LY): 183   Total luminosity (Suns): 9.9

Milburn 256	Rating: 4 D

Other name(s): **ADS 2858**

Position: **0357+6344**

	Magnitude	Separation	PA	Year	Spectra	Colors
A	10.86	—	—	—		W
B	10.54	4.7 +	275 +	1989		W

Notes:
    1906: 4.2 @ 270. Six measurements.
    Hipparcos gives discordant distances – 182 light years versus 393.
Observations:
    C-11 at 115x. Nice field.

Zone 37, Map 4

| Double Stars |

Easy

| Kruger 20 | Rating: 3 E |

Other name(s): **ADS 2573**

Position: 0331+5558

	Magnitude	Separation	PA	Year	Spectra	Colors
A	9.92	—	—	—	A2	W
B	10.32	7.5 +	297 +	1991		W

Notes:
   1890: 7.4 @ 296. Ten measurements.
   The stars have similar proper motion.

Observations:
   C-11 at 115x.

Scale model (in which the Sun is the size of a baseball):
      Diameter: A = 0.20 inches.
      Separation: AB = 2,657 feet.

   Distance (LY): 43   Total luminosity (Suns): 0.026

| STF 396 | Rating: 2 E |

Other name(s): **ADS 2592; HD 21769; SAO 24093**

Position: 0333+5846

	Magnitude	Separation	PA	Year	Spectra	Colors
A	6.45	—	—	—	A4 III	W
B	7.90	20.5 +	244 +	1991	F IV	pB
C	10.80	164.4 -	102 =	1959		No

Notes:
   AB 1829: 20.4 @ 242. Thirty-five measurements. Hipparcos/Tycho data show different distances for these stars; they may be optical. However, the stars exhibit common proper motion.
   AC 1879: 165.3 @ 102. Three measurements.

Zone 37

Observations:
C-8 at 83x. Franks saw them as W and B. Rich field.

Scale model (in which the Sun is the size of a baseball):
Diameter: A = 5.08 inches; B = 1.30 inches.
Separation: AB = 24.3 miles; AC = 195 miles.

Distance (LY): 760   Total luminosity (Suns): 152

STF 398					Rating: 4 E	

Other name(s): **ADS 2603**

Position: 0334+5817

	Magnitude	Separation	PA	Year	Spectra	Colors
A	10.30	—	—	—		?
B	10.30	9.6 -	322 -	1962		?

Notes:
1829: 9.9 @ 331. Six measurements.

Observations:
C-11 at 115x.

Webb 2					Rating: 1 E	

Other name(s): **Piazzi Smyth 97; ADS 2691; HD 22764; SAO 24169**

Position: 0343+5958

	Magnitude	Separation	PA	Year	Spectra	Colors
A	5.72	—	—	—	K5 II	D!!
B	8.70	55 -	35 +	1925	B8 II	dB!!

Notes:
1863: 55.6 @ 34. Sixteen measurements.
There is also a 13.8-m star from A at 21.4 @ 95 (1913); a 13.0-m star from A at 34.9 @ 300 (1913); and a 10.80-m star from A at 168.2 @ 161 (1909).
The primary is an infra-red source and an Alpha CVn variable.
Hipparcos/Tycho data show different distances for these stars; they may be optical.

Zone 37

Observations:
C-8 at 83x. Webb saw them as rO, B. Very rich field.

Scale model (in which the Sun is the size of a baseball):
Diameter: A = 30.0 inches; B = 1.34 inches.
Separation: AB = 189 miles.

Distance (LY): 2200   Total luminosity (Suns): 2130

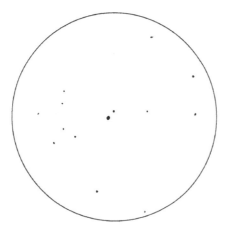

Webb 2
Double star
Observed with C11 on November 19, 2003
Magnification: 98x

South 436	Rating: 2 E

Other name(s): **HD 23594; SAO 24244**

Position: 0349+5707

	Magnitude	Separation	PA	Year	Spectra	Colors
A	6.46	—	—	—	A0 Vn	W
B	7.18	57.6 -	76 +	1992	A0	W

Zone 37

Notes:
     1823: 58.3 @ 75. Thirty-four measurements. The stars share common proper motion.

Observations:
     C-8 at 104x. Faint but rich field.

Scale model (in which the Sun is the size of a baseball):
         Diameter: A = 5.08 inches.
         Separation: AB = 47.6 miles.

Distance (LY): 530   Total luminosity (Suns): 90

Kruger 21					**Rating: 3 E**	

Position: 0352+5601

	Magnitude	Separation	PA	Year	Spectra	Colors
A	9.57	—	—	—	A0	W
B	10.14	4.2 -	279 +	1991		A

Notes:
     1890: 4.4 @ 278. Five measurements.

Observations:
     C-11 at 115x.

Scale model (in which the Sun is the size of a baseball):
         Diameter: A = 6.70 inches.
         Separation: AB = 2059 feet.

Distance (LY): 60   Total luminosity (Suns): 0.067

Moderate

Stein 1991					**Rating: 5 M**	

Position: 0331+5842

	Magnitude	Separation	PA	Year	Spectra	Colors
A	9.14	—	—	—	K	Y
B	11.13	14.2	115	1911		W?

Zone 37

Notes:
One measurement.

Observations:
C-11 at 115x.

Scale model (in which the Sun is the size of a baseball):
Diameter: A = 36.7 inches.
Separation: AB = 3.25 miles.

Distance (LY): 147   Total luminosity (Suns): 0.433

---

h2207   **Rating: 4 M**

Position: 0352+5526

	Magnitude	Separation	PA	Year	Spectra	Colors
A	9.93	—	—	—	K0 III	Y
B	11.00	10.0 +	49 +	1912		W?

Notes:
1908: 9.1 @ 43. Three measurements.

Observations:
C-11 at 115x.

Scale model (in which the Sun is the size of a baseball):
Diameter: A = 46.2 inches.
Separation: AB = 2.54 miles.

Distance (LY): 163   Total luminosity (Suns): 0.304

Zone 37

Espin 1820					Rating: 5 M

Other name(s): **ADS 2878**

Position: 0358+5712

	Magnitude	Separation	PA	Year	Spectra	Colors
A	11.84	—	—	—		W
B	12.90	5.2 +	244 -	1991		W

Notes:
   1911: 3.9 @ 256. Nine measurements.

Observations:
   C-11 at 115x.

Difficult

Engelmann 16					Rating: 5 D

Other name(s): **Bpm 48 (a, b); Hipparcos 497 (D); HD 24409; SAO 24307**

Position: 0356+5939

	Magnitude	Separation	PA	Year	Spectra	Colors
A	6.53	—	—	—	G0	yO
B	10.68	137.0 -!	40 +	1911		W
C	10.00	188.1 +	40 +	1911		W
D	9.90	0.4	321	1991		No

Notes:
   AB 1893: 153.4 @ 355. Seven measurements.
   AC 1893: 187.2 @ 39. Three measurements.
   AD One measurement.

Observations:
   C-11 at 115x.

Appendix

Zone 37

Deep Sky Objects

Moderate

King 6	**Rating: 3 M**

Position: 0328+5627

Type : Oc
Dimensions: 7'

Class: IV 2p
Cluster population: 35

Magnitude : 10.0

Notes:
  The brightest star is 10 m.

Observations:
  C-8 at 83x. Appears as a slight richening of the field; only seven stars counted at low power; high power does not help. Moderate field.

Tombaugh 5	**Rating: 5 M**

Position: 0348+5903

Type : Oc
Dimensions: 17.0'

Class: III 2m
Cluster population: 60

Magnitude : 8.4

Notes:
  The brightest star is 11.6 m.

Observations:
  C-8 at 206x. Eight stars were resolved, the brightest three forming a nice little right triangle.

Zone 37

Zone 37 Mini-Catalog

Double Stars

Star Designation	Position	Rating	Map	Component	Mag	Sep from A	PA from A
						Specifications	
Webb 2	0343+5958	1E	4	A	5.72	—	—
				B	8.70	55 -	35 +
STF 349	0311+6347	2E	1	A	7.43	—	—
				B	8.14	6.1 =	317 -
STF 362	0316+6002	2E	1	A	8.30	—	—
				B	8.80	7.3 +	137 -
				C	10.50	26.7 +	26.7 +
				D	11.10	29.9 -	29.9 -
				E	9.90	35.3 =	243 +
				F	11.00	106.6	258
				G	13.72	110.6	281
STF 373 rej	0322+6244	2E	1	A	7.63	—	—
				B	10.40	20.1 +	118 +
				C	7.78	116.2 -	116.2 -
STF 396	0333+5846	2E	4	A	6.45	—	—
				B	7.90	20.5 +	244 +
				C	10.80	164.4 -	164.4 -
South 436	0349+5707	2E	4	A	6.46	—	—
				B	7.18	57.6 -	76 +
Espin 1775	0300+5828	3E	2	A	10.24	—	—
				B	10.60	2.0 -!	288 +
STF 335	0305+6345	3E	1	A	8.61	—	—
				B	9.46	21.9 -	161 +
Scardia 9	0311+5544	3E	2	A	8.73	—	—
				B	10.36	27.8 -	137 =
				C	11.32	53.7 +	53.7 +
Kruger 20	0331+5558	3E	4	A	9.92	—	—
				B	10.32	7.5 +	297 +
Kruger 21	0352+5601	3E	4	A	9.57	—	—
				B	10.14	4.2 -	279 +
Stein 1958	0301+5627	4E	2	A	10.54	—	—
				B	11.72	14.3 +	133 -
OSS 31	0301+5940	4E	2	A	7.35	—	—
				B	8.07	73.4 -	230 +
STF 329	0301+5902	4E	2	A	8.66	—	—
				B	10.36	16.1 +	274 +
Stein 1976	0320+5850	4E	2	A	9.34	—	—
				B	12.13	13.8 -	63 -
h2185	0323+5607	4E	2	A	10.55	—	—
				B	11.60	9.3 -	249 +
Bur 1378	0327+5816	4E	2	A	8.52	—	—
				B	12.39	26.4 -	124 +
Stein 1984	0328+5627	4E	2	A	11.23	—	—
				B	12.07	11.1 -	74 -
STF 390	0330+5527	4E	2	A	5.06	—	—
				B	10.02	14.6 -	160 =
				C	10.63	110.2 +	110.2 +
STF 398	0334+5817	4E	4	A	10.30	—	—
				B	10.30	9.6 -	322 -
STF 402 rej	0336+6317	4E	3	A	7.30	—	—
				B	10.40	12.0 -	164 +
OSS 36	0340+6352	4E	3	A	6.81	—	—
				B	8.21	46.2 +	71 +
Scardia 8	0311+5605	5E	2	A	10.02	—	—
				B	11.38	23.5 +	176 -
Scardia 11	0314+5617	5E	2	A	10.74	—	—
				B	12.31	26.2	267
Stein 443	0319+5937	5E	2	A	10.72	—	—
				B	11.80	10.6	77
Scardia 14	0319+5524	5E	2	A	9.00	—	—
				B	10.49	46.9 +	293 =
STF 378	0324+5826	5E	2	A	9.55	—	—
				B	10.64	18.7 +	314 +
STF 397	0334+6024	5E	3	A	8.61	—	—
				B	9.10	5.0 -	41 -

Star Designation	Position	Rating	Map	Specifications			
				Component	Mag	Sep from A	PA from A
Ball 12				A	6.99	—	—
				B	7.10	0.2	79
	0342+6239	5E	3	C	9.66	208 =	208 =
Hussey 1063				A	9.52	—	—
				B	10.52	2.8 =	335 +
	0343+6256	5E	3	C	11.57	49.7 -	49.7 -
STF 389				A	6.14	—	—
	0330+5922	3M	2	B	7.50	2.5 -	71 +
Espin 1881				A	10.50	—	—
	0340+6310	3M	3	B	10.60	2.6 -	161 -?
Kruger 17				A	8.77	—	—
	0303+6051	4M	1	B	9.45	3.2 -	219 -
Stein 1977				A	8.20	—	—
	0321+5910	4M	2	B	10.16	5.4 +	66 -
STF 386				A	9.46	—	—
	0328+5510	4M	2	B	9.40	2.6 +	60 +
Argelander 57				A	9.59	—	—
	0341+6040	4M	3	B	9.69	6.3 +	311 -?
STF 445				A	8.56	—	—
	0351+6007	4M	3	B	9.52	3.0 =	259 +
h2207				A	9.93	—	—
	0352+5526	4M	4	B	11.00	10.0 +	49 +
OS 67				A	4.99	—	—
	0357+6107	4M	3	B	8.16	1.7 =	49 +
Stein 422				A	10.14	—	—
	0303+6412	5M	1	B	10.71	9.8 -	126 -
Milburn 254				A	10.68	—	—
	0305+6438	5M	1	B	11.90	6.6 +	192 +
P Muller 37				A	9.65	—	—
				B	11.00	0.5 =	207 +
	0310+6259	5M	1	C	12.18	17.0	17.0
Stein 1967				A	9.78	—	—
	0314+5846	5M	2	B	11.90	9.3 =	95 +
Stein 457				A	11.17	—	—
	0328+6124	5M	1	B	11.52	13.9 -	157 -
Milburn 116				A	10.34	—	—
				B	10.60	3.3 +	47 +
	0328+5625	5M	2	C	10.60	13.8 +	13.8 +
Holmes 42				A	9.40	—	—
	0330+5955	5M	2	B	10.90	5.3 =	52 +
Stein 1991				A	9.14	—	—
	0331+5842	5M	4	B	11.13	14.2	115
Stein 460				A	10.36	—	—
	0331+6402	5M	3	B	11.50	10.1	310
Stein 475				A	10.64	—	—
	0342+6216	5M	3	B	10.55	9.0 +	92 -
Espin 1820				A	11.84	—	—
	0358+5712	5M	4	B	12.90	5.2 +	244 -
A 975				A	8.44	—	—
	0309+5639	3D	2	B	10.15	1.6 =	203 +
Kruger 18				A	9.83	—	—
	0306+5744	4D	2	B	10.17	1.5 +	273 -
STF 385				A	4.33	—	—
	0329+5956	4D	2	B	7.97	2.5 +	159 -
STF 400				A	6.86	—	—
				B	7.99	1.5 =	265 -
	0335+6002	4D	3	C	10.66	92.6 +	92.6 +
Milburn 256				A	10.86	—	—
	0357+6344	4D	3	B	10.54	4.7 +	275 +
STF 384				A	8.16	—	—
				B	8.96	2 =	274 +
	0328+5954	5D	2	C	10.40	116.8 -	116.8 -
Engelmann 16				A	6.53	—	—
				B	10.68	137.0 -!	40 +
				C	10.00	188.1 +	188.1 +
	0356+5939	5D	4	D	9.90	0.4	0.4

Zone 37

Deep Sky Objects

Object	Position	Rating	Map	Specifications		Surface Br	Population
				Type	Mag		
Stock 23	0316+6002	3E	1	Oc/Gn (III 3pn)	6.5		25
King 6	0328+5627	3M	4	Oc (IV 2p)	10.0		35
Trumpler 3	0312+6315	4M	1	Oc (III 3p)	7.0		30
Collinder 34	0301+6025	5M	1	Oc (I 3p)	6.8		
Tombaugh 5	0348+5903	5M	4	Oc (III 2m)	8.4		60
IC 289	0310+6119	5D	1	Pn (4)	12.3		

Index

Adobe Acrobat: xiii, 49, 50, 51, 53, 54, 57, 59
Airy disc: 17
Aitken, Robert: 22, 23, 27
alt-azimuth mount: 1, 2
aperture: x, 2, 3, 11, 14
averted vision: 13

Baize, Paul: 24
'BARFS (big and really faint stuff): 44
binary stars: x, xiii, 23, 26, 40, 41
    See also *double stars*
binary system: see *binary stars*
Burnham, Sherburne W: 23, 24

Chester, Geoff: 46
Coe, Steve: 46, 47
collimation: 4
Couteau, Paul: 4, 24
Crayon, A. J.: xiv, 47

Dawes limit: 3 (footnote), 4
declination: ix, x, xiii, 3, 7, 8 (footnote), 9, 13, 19, 20
declination axis: 7
deep sky objects: x, xiii, xiv, 1, 2, 33, 34, 35, 37, 40, 43, 44, 56
difficulty levels: xiv, 44, 55, 56, 69
double stars: x, xiii, xiv, 1, 2, 4, 7, 17, 18, 19, 20, 21, 21 (footnote), 22, 23, 24, 25, 26, 27, 28, 29, 30, 31, 33, 37, 40, 44, 47, 55, 56, 61

equatorial mount: 2, 3, 4, 5, 7
European Space Agency (ESA): 39, 46

field of view: 8, 19, 20, 54
filters: 12, 13

galaxies: x, xiii, 2, 11, 12, 13, 14, 16, 33, 34, 37, 38, 41, 44, 46, 61
globular clusters: x, xii, xiv, 1, 13, 34, 37, 38, 39, 40, 61
Greene, Eric: 46

Heintz, Wulff: 26
Herschel, John: 21 (footnote), 26, 29, 30
Herschel, William: 26, 27
    Herschel number: 37
Hertzsprung, Ejnar: 24, 26

Jonckheere, Robert: 27

Kaler, Dr. James B.: 45, 46

light grasp: 3, 4
luminosity: 36, 40, 41

magnification: 2, 3, 9, 13, 14, 21
magnitude: x, 3, 12, 21, 21 (footnote), 22, 23, 30, 32, 33, 34, 40, 41, 43, 44, 47
Morgan-Keenan-Kellman (spectral classes): 21, 40

National Aeronautics and Space Administration (NASA): 47

objective mask: 17, 18 (illustration), 21
observatory: 5, 11, 13, 21
observing forms: 54, 67
occultation binaries: 19
open clusters: ix, x, xiii, 1, 13, 34, 36, 37, 38, 40, 46, 61
optical double: 18

planetary nebula: x, xiii, 33, 34, 37, 38, 40, 56, 61
polar alignment: 2, 8, 9
polar axis: 7, 8, 9 (footnote)
position angle: 19, 31

rating: 44
Rayleigh formula: 3
Rayleigh limit: 3 (footnote), 4
resolution: 3, 4
resolving power: see *resolution*
right ascension: ix, xiii, 3, 7, 20
Rossiter, Richard A.: 28

Saguaro Astronomy Club: 47, 54
Sawyer, Helen: 37
scale models: xii, 39, 40, 41
seeing: xiv, 11, 12, 13, 14, 16, 41, 44
separation: 19, 20, 31, 40
setting circles: 7, 8, 9
Shapley, Harlow: 37
sketch book: 13
sky album: 47, 54, 55, 67
Software Bisque: 9, 13, 45
South, James: 26, 29